Soul Power

Soul Power

✦

Science, Spirituality and the Search for the Soul

Guy McKanna

iUniverse, Inc.
New York Lincoln Shanghai

Soul Power
Science, Spirituality and the Search for the Soul

iUniverse, Inc.

For information address:
iUniverse, Inc.
2021 Pine Lake Road, Suite 100
Lincoln, NE 68512
www.iuniverse.com

ISBN: 0-595-28418-3

Printed in the United States of America

Contents

PART TWO—SPIRITUAL PHENOMENA

PART THREE—PRACTICAL SOUL POWER: AND HOW TO GET IT

INTRODUCTION

Today, many of us are looking for more from life, more than can be found at work, at home, in hobbies, or even in relationships. Some people are turning towards religion, others to new age ideas or other paths to make better sense of our lives. In fact, more of us than ever before are seeking to learn more about our souls and how to nurture our spirituality.

Yet, few of us are finding the spiritual answers we seek.

Why is this so? Other aspects of our world continue to be revealed each week, with exciting new developments in medicine, biology, astronomy and other sciences providing answers that explain the world around us and benefit our lives.

Why is this not the case with spirituality and our understanding of our souls?

Religions, once charged with protecting and nurturing our souls, are often unable to adequately answer many of today's spiritual questions, especially those following the tumultuous events of the beginning of the 21st century.

In contrast, sciences that were once the antithesis of the spiritual realm are now providing significant insights into our souls. Recent discoveries in various sciences reveal hints how our souls may operate and provide tantalizing tenets about spirituality.

But can science adequately explain such intangibles as the soul and spirituality? As a former journalist who has researched spirituality for many years the answer is a resounding "yes". Taking a multi-disciplinary approach, this book investigates and brings together revelations and evidence from history, religion, biology, chemistry, neurology, electronics, physics, cosmology and other fields to explain the nature of our souls and spirituality. (Many of the major discoveries are described by those who made them in their own words).[1]

The evidence and theories presented are designed to further discussion and objective study of our souls and spirituality. Part I investigates the science behind your soul, while Part II examines the science of several things considered spiritual and Part III provides tools and techniques based on this new scientific understanding to develop spiritual awareness. The work presented should not be considered as overriding religious traditions, rather it aims to explain the workings of our souls and raise individual spiritual understanding.

PART I—SOUL POWER

1—Defining the Soul

Just what is the "soul"?

The word means different things to different people, to different cultures. To some it is a type of food, to others a style of music, but to most people it is something intangible inside our heads that enables us to experience spirituality.

Those who report spiritual experiences describe sensing a great power, a power that often changes their lives.

Trying to determine what the soul and these spiritual experiences are is like a good detective story; there are many clues, yet little hard evidence. However, with today's modern forensic techniques, science and technology we may now be able to discover the truth. You be the judge.

The first challenge is that there is no widely accepted definition of "the soul".

Various dictionaries describe it as the underlying principle of life, the principle of thought, a seat of emotions, the spiritual part of a person, the incorporeal or non-material part of us.

There is a wider range of descriptions from various religions, even variation within the scriptures of each faith. A predominant religious description is of the soul as a sort of moral compass, distinguishing between right and wrong. The *Catholic Encyclopedia,* for instance, defines the soul as the "source of thought activity," but does not elaborate on what this thought activity is, or where its source might be. Interestingly, this encyclopedia adds "the notion that God has a supply of souls that are not any body's in particular until He infuses them into human embryos is entirely unwarranted by any evidence. The soul is created by God at the time it is infused into matter."[1] The word "spirit" has even more diverse definitions, ranging from a form of alcoholic beverage to a third element of the Christian Trinity to meaning ghost in some languages.

By contrast, new age practitioners tend to describe the soul as encompassing an individual's striving for self-transcendence, a mind within the mind, an intelligence wired to the mysteries of the universe, an awareness of something greater than yourself or an awakening as to who we are, or can be. This may or may not include religion. The *Encyclopedia of Spirituality* defines spirituality as a personal search for answers to the most profound questions of life.[2]

As for scientists, there is no generally accepted scientific definition of the soul and spirituality. Though psychologists often explain the functions of the soul as the elevated elements of the mind and thought.

The *Encyclopedia Britannica* defines soul as "in religion and philosophy, the immaterial aspect or essence of a human being, that which confers individuality and humanity, often considered to be synonymous with the mind or the self. In theology, the soul is further defined as that part of the individual which partakes of divinity and often is considered to survive the death of the body."[3]

The fact that there is no universally accepted definition of the soul or spirituality is interesting given that more than half the world's population is believed to have experienced a spiritual state. In the United States,[4] Gallup polls in the 1990s found some 53% of American adults said they had experienced a moment of sudden religious awakening or insight. The *British Medical Journal* reports 76% of people in the United Kingdom admit to having had a spiritual experience.[5] For many people, the spiritual experience is so powerful that it changes their life, forever.

For our investigation, the soul could be considered as an intangible element and process that enables us to realize and increase our awareness of our self, the world around us and of greater potentials—in short, elevated elements of mind and thought.

Just how does an intangible thing do such esoteric stuff?

There are many clues ranging from antiquity to today.

People and societies from prehistoric times onwards have expressed surprisingly similar concepts about the soul. For thousands of years Chinese peoples have referred to the soul as a "vital breath" called "chi" or "qi" that is believed to be an energy coursing through our bodies and the universe. The Japanese call it "ki". In ancient Greece, there was a similar tradition of "life breath" known as "pneuma". In ancient Rome, the Latin word "spiritus" also meant breath of life and provides the basis for the English words spirit and spirituality. Tibetans refer to spirit as a subtle natural energy called "lung". In India's Sanskrit, there is an energetic breath called Prana, while Muslims call this life force Barraka. Lakota Sioux Indians in North America call it Neyatoneyah, while the busmen of the Kalahari Desert in Africa refer to it as Num, which means boiling energy. All these references suggest the soul and spirituality is something akin to an energetic breath, breath of energy, or some form of energy.

In fact, most early religions, with the exception of Christianity, generally conceived of and described God as an immaterial, incorporeal being.[6] It has only been comparatively more recently that God has been anthropomorphosized and given physical human attributes.

Another clue is found in the many ancient myths about the soul and spirituality, which are similar across cultures, countries and time. Mythologist, Joseph Campbell studied the world's legends and religions and detailed their many common elements in his various books. He noted that, typically, the mythic hero sets forth from his hut or castle, is lured, kidnapped or voluntarily proceeds on an adventure where he (or she) encounters a shadowy figure that guards the solution to the hero's quest. While attempting to pass this threshold, the hero is threatened by strange beings and occasionally aided by them. Ultimately, the hero undergoes a supreme ordeal and gains a reward of spiritual growth. The experience profoundly changes the hero, with an expansion of consciousness that in turn makes him or her a good leader and in turn helps their community.[7] Campbell believed this striking similarity of myths across cultures, countries and times suggests there is a common core to the expression of human spirituality.

He also noted that many cultures have also expressed that there is some kind of order or authority to the universe we inhabit. Not knowing just what this authority is, people have postulated deities of many kinds that exercise spiritual authority over the material and mechanistic world around us.

Psychologist Carl Jung took the commonality of these myths and legends a step further and suggested that if a concept is in the psyche of a large number of people it must be there for some reason.[8]

But just what is the purpose of having a soul that expresses spiritual experiences?

Several sociologists and scientists believe we have evolved a spiritual aspect to our minds as a survival strategy, as organized religion reduces anxiety, promotes compassion and encourages societies that protect people.[9] The brain is genetically programed to encourage religious beliefs, suggests Andrew Newberg and Eugene d'Aquili, who say spirituality is a necessary part of human survival.[10] They say spirituality allows us to interpret random happenings as part of a larger and more holistic reality, which allows us to avoid emotional paralysis and cope with anxiety. Others suggest spirituality developed to better the human species and talk of concepts such as the "evolution of a global consciousness" and the like.[11]

However, this may not necessarily be the case, as history records how irrational and dangerous religiousity can be to society, with people using gods as justification to kill others.

Further research is obviously required for a more definitive answer. The only problem is that the more research you undertake on the history of the human soul and spirituality, the more you realize that the issues are the same ones that existed 4,000 years ago. We haven't come very far spiritually.

Moving on from the controversies and confabulations of history to today, to describe someone as spiritual and someone else as not is to describe their differing awareness and response to things spiritual. For instance, some people readily accept what a religion suggests the soul and spirituality are, while others have to determine what it is for themselves. This is one of the major differences between Eastern (Asian) and Western spirituality, with the former suggesting it comes from within ourselves and the latter suggesting it is bestowed from outside.

Today in the West, the soul is most often described in holistic discussions on human health, where it is generally portrayed as one side of an equilateral triangle, with the other sides being the body and the mind. It is suggested that all sides should be equal, otherwise the triangle and a person's overall health or life is weakened. This modern description extends a view originating in antiquity, where people saw no distinction between body, mind and spirit and that physical, emotional and spiritual aspects of life all needed to be in balance.

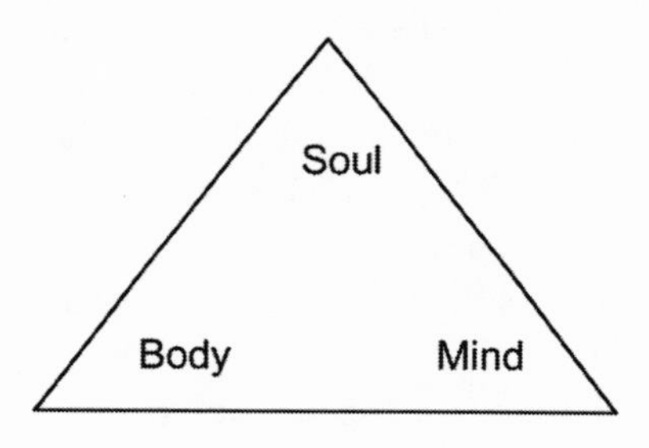

Figure 1. Body, mind, soul triangle

Another modern representation places the soul at the top of a pyramid, with the bottom section being the physical body, the middle component being the mind—both capped by a soul pointing to the heavens. The famous psychologist Sigmund Freud suggested a similar pyramid, with the conscious ego, topped by an unconscious super ego and then the Id. These representations do not define the soul or its location; rather they suggest how it is related to the mind and body.

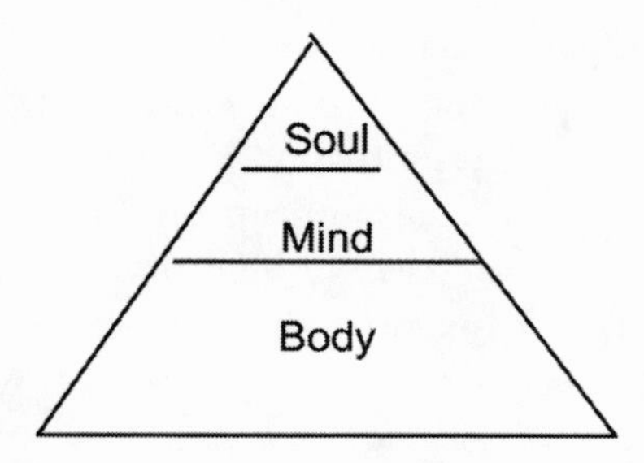

Figure 2. Body, mind, soul pyramid

If you ask a person on the street about the soul and things spiritual you are likely to get an answer that comprises some element of:

- love, knowledge, compassion,
- an elevated component of mind or consciousness,
- individuals being part of a greater whole,
- some recognition of a presence of a benevolent higher power,
- a seeking of answers to questions such as Why am I here? Where am I going? and/or
- a selfless, holistic approach to life.[12]

This striving for knowing more, for a reason for our existence, seeking a higher realm outside ourselves and that influences who we are is a hallmark of today's human spirituality and what continues to draw people to church, to prayer, to meditation and to seek those powerful spiritual experiences.

Yet, it remains intriguing how something referred to by most people at some stage in their life, has not been adequately explained before. It is a great irony of spirituality that the only species on earth apparently capable of such experiences understands so little about them.

Some people claim no definition of "soul" is required as there is no such thing. They make a very logical case that we are like the android robots of science fiction, programed with a spiritual delusion to make us believe that our lives our are worthwhile and overshadow a meaningless existence and inevitable death.[13]

In an interesting aside, higher levels of education have been found to be associated with lower levels of religiosity. Studies suggest that of all the variables influencing peoples' religious attitudes, education is one of the most important and most powerful.[14] Yet, as some people become even more highly educated they seek more answers regarding spirituality. This would suggest that knowledge and awareness could be a component of spirituality.

However, just because we can't see the soul does not mean that it does not exist. Our world is full of such apparent conundrums. For instance, while a person's character cannot be easily described or its location pinpointed in scientific terms it does not mean that it does not exist and does not impact others. Similarly, take a photon, the element of light that has no charge or mass yet has been scientifically proven to exist.

To get some practical understanding of your own life force or "chi", try this basic exercise taught in Chinese medicine. Stop, relax, close your eyes and slowly bring both your hands together with palms open and facing each other. Before your hands touch, you may feel some slight resistance or slight increase in heat, sensation or pressure.

This effect is supposed to reflect your "life force" or spiritual energy. The only problem with this is trying to verify it scientifically, as the effect varies for every person. Is it truly an indication of our spiritual energy or is it merely a biochemical effect of raised arms pumping out more chemicals?

This is the same problem with many things described as spiritual today. For example, the bending of spoons has nothing to do with spiritual or mental powers, but a lot to do with physics and preparing the spoons beforehand. The same applies to magicians' tricks, which can all be explained logically and scientifically, even though they may appear mystical.

So what can logic and modern science reveal about the soul and spirituality?

Science has already rendered some of the major religious issues that occupied theology scholars over past centuries meaningless in a physical sense, such as the whereabouts of heaven and hell. For example, advances in astronomy have shown that heaven is not up above us, in the universe.

Surgeons have not detected anything physical that we could call a soul in the human body. This is despite some philosophers and early doctors suggesting that the pineal gland the brain was the source of the soul. Removal of your pineal gland does not remove your spirituality.

These discoveries have prompted many theologians to reinterpret the original symbolism beyond literal meanings.

What is known for certain is that the ability to express spiritual insights requires a body and a mind. Jesus Christ, Muhammad and the Dalai Lama first thought, consciously sensed and conveyed their spirituality while alive.

2—Chemical Bodies—Chemical Souls?

Can what we know about our physical bodies tell us anything about our immaterial souls?

At their most fundamental level, our bodies are believed to be comprised of the basic building block of the universe, minute "strings" of energy. These strings are joined together to form quarks, which in turn form three subatomic particles—electrons, protons and neutrons. Protons and neutrons join together to form the dense central core of atoms. Protons are positively charged while neutrons do not have any electric charge. Negatively charged electrons move about the space around the central nucleus of the atom.

Atoms are extremely small, with more than 200,000 of them fitting into the full stop at the end of this sentence. Atoms are the smallest units of matter that can form chemicals, create compounds and create chemical reactions. Atoms, such as calcium, carbon, hydrogen, nitrogen and oxygen are essential for maintaining life on earth. The human body is composed of 25 atomic elements.

Oxygen	65%	8 electrons
Carbon	18.5%	6
Hydrogen	9.5%	1
Nitrogen	3.2%	7
Calcium	1.5%	20
Phosphorus	1.0%	15
Potassium	0.4%	19
Sulfur	0.3%	16
Chlorine	0.2%	17
Magnesium	0.1%	12
Iodine	0.1%	53
Iron	0.1%	26
Boron, Fluorine, Aluminum, Silicon, Vanadium, Chromium, Manganese, Cobalt, Copper, Zinc, Selenium, Molybdenum and Tin	<0.1% [1]	

Figure 3. Chemical composition of the human body

When these atomic chemicals join together they form molecules. When certain molecules combine together they can form amino acids, proteins, DNA and other structures that can form cells. Cells are the basic functional units of an organism and are the smallest living units in the human body. There are many kinds of cells in the human body, such as blood cells, muscle cells, nerve cells and so on.

The next level of structural organization in our bodies is the tissue level. Tissues are groups of cells and the materials surrounding them that work together to perform a specific function. There are four basic tissue types in the human body—connective tissue, epithelial tissue, muscle tissue and nervous tissue.

When different types of tissue are joined together they form organs, which have specific functions. Organs include the brain, heart, liver, lungs, skin, stomach and so on.

The next level of the body's structure is the system level. This comprises organs that have a common function, such as the respiratory system or the digestive system. Sometimes an organ is part of more than one system.

Starting with this basic scientific understanding of the human body, which level does the soul operate at: the atomic, cellular, tissue, organ or system level?

We know it is not at the tissue or organ level, as doctors and scientists have not discovered a soul in all their probing and slicing into the human body over the past millennium.

Some research suggests that it could be at the chemical or molecular level. For instance, in one famous experiment, chemicals were zapped with electricity to form the basic building blocks of life. Gases believed to have been present in the earth's early atmosphere were mixed together in a bottle and charged with electricity by Stanley Miller in 1953, creating amino acids—crucial constituents of life. Similarly, scientists Frank Cole and Ernest Graf theorized that as the earth developed, micropulsations in the planet's magnetic field created lightning and electrical currents that mixed together chemicals to form biological molecules, such as proteins, which could have eventually formed primitive cells and ultimately evolved into living organisms.

In another interesting development, scientists have discovered primitive forms of life on asteroids and meteors—and suggested that life has spread throughout the universe this way. It is believed these dormant cells awaken from their cold slumber when their rocky space ship strikes a suitable planet and warms them up, out of their slumber. This is given some credence by that fact that we are composed of the same chemical elements that are found throughout the universe, that we are made of the same stuff as stars.

These findings certainly encompass the "energetic breath" element of life and the soul as described in antiquity. In fact, there is unequivocal evidence that wherever there is life, there are electric properties: dead matter has a very different electromagnetic potential compared to identical living matter, despite the same chemical composition.[2] Could the soul then be a mix of chemicals and the appropriate "cooking" temperature?

Science suggests not, as while our spirituality may seem real to us, the atoms and molecules of which we are composed are impartial and unthinking. Cellular processes in the body are passive and mechanical, simply doing what they are directed to do. In other words, cells and atoms simply follow the direction of the brain and nervous system, which follows the laws of biochemistry, which follow the rules of chemistry, which in turn obey the laws of physics.

However, this is not necessarily as straightforward a situation as it might seem, with scientists recently finding that impartial molecules can convey emotions. Scientists recognize that some molecules in our bodies play an important role in terms of our emotions—emotions that help keep us alive, protect our physical and mental selves and encourage us to reproduce our kind as well as make our lives emotionally richer.

Chemical molecules, such as neuoropeptides and associated receptors, form a mind-body or "psychosomatic communication network," says Candace Pert, pharmacologist and professor at Georgetown University and formerly chief of brain chemistry at the U.S. National Institute of Health.

Pert found that peptides convey emotions throughout the whole body, not just the brain. "The peptide network expends beyond the hippocampus, to organs, tissue, skin, muscle and endocrine glands. They all have peptides receptors on them and can access and store emotional information. This means this emotional memory is stored in many places in the body, not just the brain."[3]

Through this peptide network, she says we can access different memories, mood states or developmental stages. "You can access emotional memory anywhere in the peptide/receptor network, in any number of ways. For example, if you have a memory that has to do with food and eating, you might access it by the nerves hooked up to the pancreas. You can access through any nodal point in the neural loop. Nodal points are places where there is a lot of convergent information with many different peptide receptors. In these nodal points there is potential for emotional regulation and conditioning."[3]

Pert suggests emotions are cellular signals that are involved in translating information into physical reality, literally transforming mind into matter. They operate at the nexus between matter and mind, interacting back and forth between the two and influencing both.

As a result, Pert believes our bodies can be a battlefield for war games of the mind. All unresolved thoughts, emotions and negativity shows up in the mind and the body and can make us sick. This leads to her theory about emotions, in particular what happens to those that are not expressed. "I believe that emotion is not fully expressed until it reaches consciousness. When I speak of consciousness, I include the entire body. I believe that unexpressed emotion is in process of traveling up the neural access. By traveling, I mean coming from the periphery, up the spinal cord, up into the brain. When emotion moves up, it can be expressed. It takes a certain amount of energy from our bodies to keep the emotion unexpressed. There are inhibitory chemicals and impulses that function to keep the emotion and information down. I think unexpressed emotions are literally lodged lower in the body. In my mind, there are levels of integration. You are integrating lower brain areas when you move the emotion up and get it into consciousness. That's where you begin comprehension. I often tell a story in my lectures. I show a picture of a woman with hot coffee, who has dropped the cup and burned herself. She reacts to the scalding coffee by being startled and feeling pain. The emotional reflex moves up and up and up the body. When it finally gets to the level of the thalamus she says, 'Oh, it's hotter than it usually is.' But then I make a joke. I say, 'It's only when it gets all the way up to the cortex that she can actually blame her husband.' That's where we put the whole spin on it. Unexpressed emotions are buried in the body—way, deep down in the circuitry of the organs, or the GI tract, or a loop in a ganglium. We even know what the memory storage looks like. It's protein molecules coupled up to receptors. Some thought it only gets stored in the brain. But it looks like that in the body, too. Your memories can get stored that way in a pancreas, for example."[4]

Pert also says there is evidence that unexpressed emotion causes illness. "The raw emotion is working to be expressed in the body. It's always moving up the neural access. Up the chakras, if you will, but really up the spinal chord. The need to resist it is coming from the cortex. All the brain's rationalizations are pushing the energy down. The cortex resistance is an attempt to prevent overload. It's stingy about what information is allowed up into the cortex. It's always a struggle in the body. The real, true emotions that need to be expressed are in the body, trying to move up and be expressed and thereby integrated. That's why I believe psychoanalysis in a vacuum doesn't work. You are spending all your time in your cortex, rather than in your body. You are adding to the resistance."[4]

Accordingly, minute physiological changes at the cell level can translate into large changes in behavior, attitude, emotions and more.

Pert notes the autonomic nervous system is also pivotal to how we learn and remember, noting that every peptide that she ever mapped could be found in the autonomic nervous system. Many molecules once thought to be only in the brain are now known to be present throughout the body. And vice versa, chemicals once thought to be only made in the body are now known to also be made in the brain. For example, insulin, which was supposed to be produced only in the pancreas, is also made inside our heads.[5]

These discoveries by Pert and others shows how the brain and body is more connected than was previously thought, with one influencing the other and vice versa—and how emotions can affect, even alter, our biology. And if emotions are able to alter biology, then surely they can affect our minds, our personality and even spirituality.

Pert speculates the mind is the flow of information as it moves among cells, organs and bodily systems, and that while it is immaterial it has a physical substrate of both the body and the brain. This "pyschosomatic information network" links the nonmaterial mind, emotion and soul, to the material world of molecules, cells and organs. She suggests this "information mind" links and coordinates cells, organs and major systems in an orchestrated symphony of life. It is this information running the systems that creates behaviour. As a result, mind becomes matter, with consciousness creating reality.

Interestingly, Pert believes happiness is when our biochemicals of emotion are open and flowing freely, integrating and coordinating our cells, organs and systemsin a smooth and rhythmic movement.[6] She suggests the oneness of all life is based on the fact that our molecules of emotions are all vibrating together.

Is a similar process involved in creating spiritual experiences, are our souls a soup of chemicals and molecules?

Some people would say yes, that our souls and spiritual experiences are somehow derived from chemicals and molecules. For example, following Pert's revelations some scientists and reporters suggested the angels be renamed serotonin and dopamine, after these chemical neurotransmitters.[7]

But while chemicals, molecules and a spark of electricity might be able to form the basic building blocks of life, can they account for the soul or spirituality? Are there also be molecules of spirituality?

The evidence so far suggests there are other factors involved. For instance, there is the question of how do these molecules of emotion know where to go; why do some of them go to distant parts of the body, while others to the cell next door, while others decide to stop mid-stream and form "emotional blockages"? There appears to be no innate intelligence inside them directing them what to do.

Scientists have also found some of these molecules in organisms that appear to have nothing remotely resembling a soul. For example, insulin has been found to be generated in one-celled organisms, which appear unable to express any emotion, let alone have any soul.[8] This suggests that peptides, while possibly involved in the functioning of the soul, do not alone account for spirituality.

Also, around 95% of the atoms in your body were not there 12 months ago. The atoms and cells in your skin cells are renewed each month, while the atomic elements of your bony skeleton are refurbished every three or more months and every seven years every cell in our bodies is believed to have been replaced. Yet, we retain our identities and appearance well beyond this time, suggest that the molecules and cells themselves do not contain our souls.

There seems to be something else beyond chemicals and molecules at work here providing direction and continuity. Is it the electric forces that underlie each and every chemical action?

3—The Force of a Nervous Soul

There have been some intriguing discoveries recently by neurologists, the scientists who study the nerve cells in our brains, that offer tantalizing clues as to the nature of our souls and spirituality. To understand these discoveries, we have to understand the whole human nervous system, which is divided into various branches and subsets.

- The central nervous system consists of the brain and the spinal cord and correlates incoming sensory information and is regarded as the source of thoughts, emotions, learning and memories,
- The peripheral nervous system which comprises all nervous tissue outside the central nervous system and includes cranial nerves, spinal nerves, ganglia and sensory receptors. The peripheral nervous system is further subdivided into the:
    - Somatic nervous system (under voluntary control),
    - Enteric (stomach) and other organ nervous systems, and
    - Autonomic nervous systems. The autonomic nervous system keeps our bodies functioning somewhat automatically and unconsciously. It is further divided into the:
        - Sympathetic nervous system, which controls organs in times of stress, and the
        - Parasympathetic system, which controls organs when body is at rest.[1]

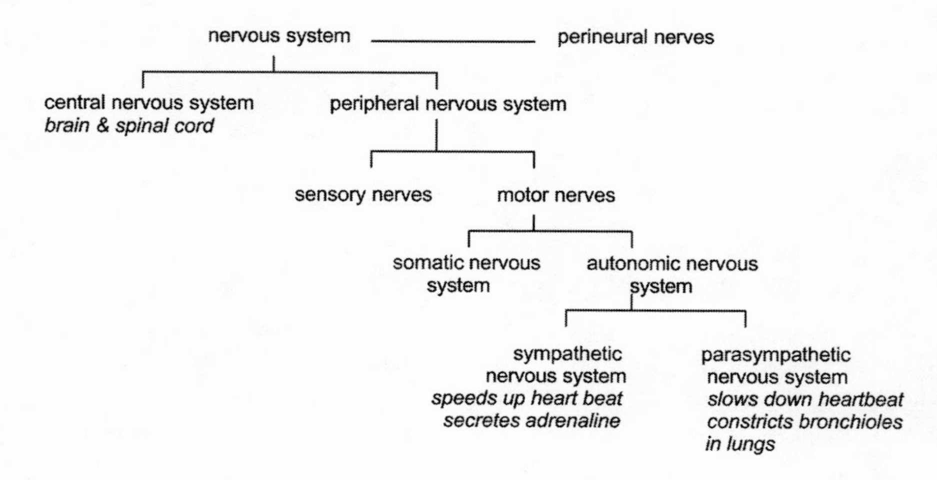

Figure 4. The human nervous system

There are different types of nerve cells. Afferent or sensory peripheral nerves collect information from the body's senses and transmit it toward the central nervous system. In exchange, efferent or motor nerves transmit information away from the brain and spinal cord and make the body (through glands and muscles) work. Interneurons, in the central nervous system, connect sensory and motor nerves and carry impulses between them.

Nerve cells (known as neurons when located in the brain) are a little different to most other cells in our bodies. While most cells measure a few millionths of an inch across, neurons have filaments projecting from them that can extend several feet (such as from the brain down the spine to the legs) or just a few microns to the neuron nextdoor. These projections enable nerve cells to send or gather information over long distances. These projections are called axons if they transmit messages and dendrites if they receive them.

Nerves are concentrated in, but not restricted to, our brains. While it weights only three pounds and accounts for just 2% of the weight of an average person, our brains use between 20–25% of the oxygen we consume as well as significant amount of the calories we eat. It is estimated that 100 billion nerves in a brain generate about 20 watts of power at any instance—enough to power a small light bulb!

Nerve cells convey messages throughout our bodies via electric waves called action potentials. This is an on-off system, much like that of a digital computer with its 0—1 coding and processing. This current is different to the alternating current (AC) in the home, where the electric field collapses and reappears with its poles reversed each time the current changes or alternates direction.

Unlike the electric current in a metal wire or power cord, which is caused by the movement of electrons moving along the wire, action potentials occur when electrically charged atoms, known as ions, of sodium and potassium move in, along and out a nerve cell—in short, a wave of energy that flows down the cell.

Messages flow from one nerve cell to the next at junctions called synapses, where they may or may not trigger a chemical messenger, a neurotransmitter, to alert the next nerve cell, where special receptor proteins wait for the correct neurotransmitter to excite them and fit into their receptors.

Depending upon how many receptors of a nerve cell are stimulated, the next nerve cell may become excited and fire energy pulses of its own, or it may be inhibited, dampening down any firing that is already occurring.

To better understand just how this energy flow works, we first have to better understand electric current.

Electricity is one aspect of one of the four fundamental forces in our world, in fact throughout the universe. Physicists believe all the interactions between various objects and materials in our daily lives can be reduced to combinations of these forces:

- electromagnetic force,
- the weak force,
- the strong force, and
- gravitational force.

Electromagnetism is the force that produces electricity as well as magnetism. Besides powering many of our modern conveniences, it occurs naturally and ranges in strength from violent lightning storms to gentle sensations of touch. Under the electromagnetic force, one negatively charged electron repels another negatively charged electron. In contrast, a positive and a negative charged electron are attracted, as in a magnet attracting iron.

The second force, the weak nuclear force, affects high-energy particles. When these particles are slowed this leads to decay and increase in mass of the particles, such as radioactive decay of uranium.[2]

The third or strong nuclear force holds protons and neutrons together in the nucleus of atoms, while at the same time stops electrons from collapsing into the center of an atom. It is believed to only operate at very short distances.

In contrast, the fourth force of gravity is a wide-ranging force of attraction. Every particle, everywhere feels the force of gravity according to its mass or energy. This is despite gravity being the weakest of the four known forces. But as it acts over large distances it adds up to produce a significant force. We all know from first-hand experience about gravity, how the earth orbits the sun due to the effects of gravity, how it keeps us flying off our small blue planet as it spins and how it causes apples to fall—and how one struck Isaac Newton's head in the late 1600s. What is not as well known is that all objects exert gravitational attraction to each other. The pull of this weak force between objects increases the closer they are, and if one object comes twice as close to another the attraction is squared. So why doesn't this page rise to your fingers as they get closer to it? It does, but only when your finger is very, very close. While a weak force, gravity can be very strong under certain conditions. For example, in 1916 Einstein in his theory of general relativity discovered that gravity's force even acted on light, bending it. He ultimately explained gravity as the curvature of four-dimensional space-time caused by matter.

At present, we are unable to explain why our world, why our universe is composed of these four fundamental forces and associated particles. (This is despite there being a lot more known about these forces than outlined above, but this is enough to demonstrate that they could be involved in an "immaterial" or "intangible" soul).

The most important force for our investigation is the electromagnetic force, with several chemicals and molecules in our bodies having negative or positive electromagnetic charges and the transfer between them of these electrons and their charges powering many physical processes.

When atoms add or loose an electron, they are known as ions, and can have a negative (-/anion) or positive (+/cation) charge.

Quantum physics explains why these charged electrons repel or attract other charged electrons—or why the south pole of a magnet repels the south pole or attracts a north pole of another magnet. The theory is that electrons, circling around the atom's central nucleus much like the moon around the earth, give off exchange particles and it is these particles that actually reach out and repel the other electrons at a distance.[3]

These exchange particles are called photons—the same photons that comprise visible light! These photons are a little like a tennis ball between two players, dart-ing back and forward between electrons telling them what to do, influencing one then the other. Sometimes this interaction can set up a resonance or wave, like a rally between tennis players.

Interestingly, while an electron is the kind of particle that physicists think of being associated with matter, a photon is thought of as being associated with a wave of light. (Remember this, as it is important and we will come back to it later).

This process is what makes seemingly inert chemicals act as directed by another force. This universal force of electromagnetism underlies all chemistry.[4] It is what allows atoms to join together to form molecules. For instance, H_2O is a molecule of water that has two hydrogen atoms and one oxygen atom that shares two electrons with the atoms of hydrogen. Each atom of each element has characteristic ways of sharing their electrons/ions when interacting with other atoms.

These ions play a major role in our cells, in our lives. Each of our cells has a plasma membrane that selectively allows some substances to enter or leave the cell. The membranes sort positively charged ions on one side and negative ones on the other (generally the inside). This polarizes cells and creates an electrical gradient between the negative and positive ions, which is insulated by a lipid layer in the center of the membrane. In short, this creates a cellular battery.

In this cellular battery the flow of ions constitutes the current that powers many of varied functions in our bodies. This is aided by dissolved bodily fluids called electrolytes, which can also conduct electric current.

The molecules that most often transfer energy from one cell to another in humans is adenosine triphosphate (ATP) which is sometimes called the "energy currency" of cells. ATP is crucial in the chemical reaction that breaks down complex molecules such as food into simpler ones and releases energy to power our bodies.

Positively charged hydrogen ions are a key element in the ATP energy process. These ions are so important that the well-known term pH is used to describe the concentration of them in fluids. The maintenance of the correct pH balance is important as the shape of proteins, which enables them to perform specific functions, is very sensitive to changes in pH. If our bodies are unable to dispose of H^+, acid-levels levels rise and quickly lead to death. This is one reason why water (H_2O) is so important to us, and makes up between 55–60% of our body mass. The rapid movement of hydrogen ions in and out of cells creates electricity that is used by ATP to ultimately pass electric impulses along nerve cells. This was first identified in 1952 by Alan Hodgkin and Andrew Huxley, who won a Nobel Prize in 1963 for their work.

But it was not until 1998 that the structure of the ion transfer channel in cell membranes was revealed by a team at Rockefeller University. Potassium ion channels underline all our movements and thoughts, says the university's Rod MacKinnon. "They are present in the lowliest amoebae and in the most complex cells, such as the human brain. In humans, the width of the potassium ion channel is 100,000 times thinner than a sheet of paper (at 6 Angstroms wide)."[5]

Interestingly, besides these voltage-operated ion channels in cell membranes, there are also other ion channels that are operated mechanically. They are believed to open or close in response to vibrational stimulation, such as touch, stretching or even sound waves.[6] Could this explain the effectiveness of new age practices such as sound therapy (sound waves), yoga (stretching) and touch therapies?

In nerve cells, the membrane contains different types of ion channels that open and close in response to different types of stimulation.[7]

Could this ionic energy be the power source of a non-material soul?

Possibly, even most probably. The result of ion exchange is certainly an immaterial energy that powers the body and creates life as we know it. This power could also accurately be described as a "life force" or "life energy".

It certainly provides a power source for our nerves, which include those in the brain, where the soul has traditionally been thought to be located.

All this would suggest that we are on the right path in terms of our investigation. However, something still seems to be missing, as other animals have similar ion channels, cellular and nervous systems, but do not exhibit what we would call a soul. Accordingly, there seems to be more to our soul and spirituality than just these cellular batteries and having a nervous system.

Maybe it is the vast extent and complexity of the human nervous system. Many philosophers, psychologists and scientists today consider the soul to be related to the development of our complex mind and memory. Practical experience certainly suggests that our soul and spirituality certainly appear to be influenced by learning and memories. Can the way we learn and think reveall more about our souls?

4—Good Vibrations

At present, biologists believe we perceive, learn and remember things due electrical actions in the nervous system, in particular the synchronized electric activity of nerve cells.[1]

As soon as your eyes see an object, different cells and nerves register the objects' horizontal, vertical and curved contours, colors, contrasts and more. Through synchronization, these various electrical impulses from different and diverse sensations are aligned and combined. Ultimately the brain registers them as the objects that you are seeing, such as the words on this page. Information from other senses such as touch, taste, sound or smell are also added and emotional elements tagged to these impulses if appropriate.[2]

Before learning, each neuron responds similarly to each stimulus. After learning, the synapses between the neurons have been slightly modified so that they react differently to different stimuli and thereby generate different memories, reactions and individual responses.

Learning and memories are believed to be created by resonating electric waves in circuits of nerves in our brain. Something stimulates one nerve, which in turn stimulates another, which stimulates another and so on until it repeats back through the circuit over and over. Neurons become more responsive to one another, so responsive they communicate even when they are no longer being stimulated by the external source.

This electric reverberation or resonance is believed to last only for a few seconds for short-term or "working" memory. In contrast, longer-term memories are encoded by longer resonance. Electrical reverberation at some neuron synapses within the hippocampus are enhanced for hours or even weeks after a brief period of high-frequency stimulation.

This important process is called long-term potentiation or LTP.

Scientists have determined that LTP involves a strong depolarization of the postsynaptic membrane of nerve cells that leads to ion exchange, as mentioned earlier. In short, nerves cell are depolarized or switched from positive to negative—basically, from off to on. These resonating cells can synchronize their firings with great precision, within milliseconds of each other and can generate a current as high as 6 milliVolts.

You can consider LTP as a little like a laser. Lasers (Light Amplification by Stimulated Emmission of Radiation) amplify light by refocusing it back onto itself further stimulating electrons, which in turm stimulate even more electrons to reverberate in unison, in turn radiating outwards in a beam of intensely focused light. Another anology is that of a musical note: while one piano key makes a note, sequences of notes that are combined,run together and overlap can make music. This appears to be what happens in the brain with neurons engaged in LTP reverberating coherently and restimulating themselves and others, with memory the result.

Neurologists have found that this process in the human brain occurs in neurons in the amygdala, hippocampus and thalamus as well as "chattering nerve cells" within the cerebral cortex.

The chattering cells in the cerebral cortex are a group of excitatory pyramidal cells that are common to certain layers of the cortex and appear to be grown just to resonate. They are known to connect with, and propogate resonating rhythms to, other cells both nearby and farther away in the brain.[3] Interestingly, they show no sign of adaption, a process where many cells adapt to what's happening and stop doing it. For instance, non-rhythmic neurons will spike in unison and resonate until the stimulus becomes very strong and then tend adapt to it and cease firing in resonance. These chattering pyramidal neurons are also known to be affected by stochastic resonance, which can help them to continue to fire.[4] (More on this particular type of resonance later.)

Neurologist and author Joseph LeDoux suggests information from the outside world reaches the resonating cells in the amygdala via direct pathways from the thalamus to the amygdala, or what he calls the "low road," as well as by a more circuitous route through the cortex, or the higher resolution road. While the low road is shorter and faster, it does not provide the amygdala with the higher resolution that is achieved via the high road, where the cortex more accurately represents the perception through more connections. This way, the brain receives two sets of messaging to build up a more accurate picture and response, he suggests.[5]

As more and more cells resonate at the same electric frequency, they set-up electric waves and fields that echo to other non-LTPing nerve cells, even to other types of cells in our bodies.[6] For instance, LTP is also known to take place in the spinal cord.[7] This is an important point, as it demonstrates that it is not just the brain where learning can occur, but also other parts of the body!

In their theory of nerve resonance called electroconformational coupling, neurologists James Weaver and Dean Astumian suggests cell membranes also use energy from other electric fields as well as fluid around them to drive chemical reactions. "Our research stresses that the membrane electric field is very dynamic—oscillating and fluctuating. Cells use dynamic electric fields to accomplish the necessary functioning of life."

They also note that proteins change shape in response to different electric frequencies.

Proteins and enzyme also absorb and store electric energy when they change shape. "The energy absorbed from the electric field is not just dissipated as heat. An enzyme, for instance, can use it to accomplish valuable functions such as transporting nutrients into the cell against a concentration gradient, forming ATP for the cell to store for later use, and signaling to the inside what is going on outside," Weaver and Astumian say.[8]

This can also re-ignite the ion exchange and restart the process, with resonating electric waves helping to bind ions to receptors on the surface of cells.[9] For instance, once cells start resonating in LTP, the neurotransmitter glutamate is produced and in turn reacts with NMDA (N-methyl D-aspartate) glutamate receptors on postsynaptic neurons.[10] This causes the NMDA receptor to open its ion channel and allow calcium to enter the postsynaptic neuron. The reaction between glutamate and NMDA stimulates the production of a protein called kinases, which also open up the ion passages in cell membranes. This also manufactures a messenger molecule of nitric oxide, which in turn flows back to the nerve axon and stimulates more glutamate production, keeping the whole process going and the energy flowing and repeating. Interestingly, NMDA is receptive to glutamate only if enough energy is passing along the nerve cell. Similarly, the flow of sodium ions is influenced by calcium and potassium ionic currents.[11] And the binding of calcium to calmodulin is enhanced by electric waves, which have also been identified as helping to increase bone cells' production of insulin-like growth factor 11.[12] In another study, pH was also found to be important to nerve cell resonance. Raising the pH of intracellular fluid by 5.5% was found to trigger the nerve cells of giant squids to fire automatically without any other stimulation.[13]

In short, once the energetic resonance of LTP is initiated it can prompt a whole electrochemical action. A detailed description and understanding of how the chemical aspects of LTP works is contained in LeDoux's benchmark book, *Synaptic Self*.[14] He says that synapses are also tagged with proteins during learning, and that only those synapses that have the associated protein tag are able to use the new protein and new memory.

Chemicals are obviously important to the overall process. However, some researchers such as Stuart Hameroff note that only about 15% of axonal action potentials reaching pre-synaptic terminals result in actual release of neurotransmitters. And with proteins and molecules targeting their receptors, and not others, at incredible speed, it appears there is also some directing element. Otherwise life would be a jumble of chemical reactions that would be unlikely to reach the higher complex order that is life as we know it.

LeDoux admits, "Mental states are not represented by molecules alone, or even by a mix of molecules. They are instead accounted for by intricate patterns of information processing within and between synaptically connected neural circuits. Chemicals participate in the synaptic transmission, and in the regulation or modulation of transmission, but it is the pattern of transmission in circuits, more than the particular chemicals involved, that determines the mental state."

James Oschman in his book *Energy Healing: The Scientific Basis*[15] takes this a step further and says in the past, the words of the 'language of life' were thought of as nerve impulses and molecules, "but we now see that there is a deeper layer of communication underlying these familiar processes. Beneath the relatively slow moving action potentials and billiard ball interactions of molecules lies a much faster and subtle realm of interactions. This dimension is subatomic, energetic, electromagnetic and wave-like in character. The chemical messenger ultimately transfers its information electromagnetically. Hence the electromagentic code is actually primary. Nerve impulses and chemical messengers are contained within the individual whereas energy fields radiate indefinitely into space and therefore affect others who are nearby," says Oschman.

This fundamental electromagnetic direction also produces another major development. As while LTP can preserve memories for hours or days, there is also an advanced stage of the process where permanent memory storage is created. The LTP resonance and chemical dance ultimately alter nerve cells, in particular the synapses between them to change shape, and create new circuits that can remember things over the longer-term. Studies have proven that learning causes changes in the neurological wiring of sea slugs among other animals, a discovery for which Eric Kandel won a Nobel Prize in 2000.

And if LTP and learning changes the electrical circuits of a sea slug, imagine what it can do to a human brain. Electron micrographs of human nerves subjected to prolonged electrical activity show a structural change, with an increase in the number of presynaptic terminals, enlargement of presynaptic bulbs and increased number of dendrite branches in postsynaptic neurons.

Research suggests the altering of some synapses occurs during sleep, with the brain using this down time to process information obtained during the day into more permanent memories.[16] This may be during rapid eye movement (REM) sleep, as it has been demonstrated that disrupting REM sleep also disrupts the formation of long-term memories. In sleep, the brain changes and can become a lot more synchronized that when you are awake. Instead of different parts of the brain firing, neuron communications appear to become much more coherent. Some scientists believe that this coherent state during sleep encourages internal dialogues within the brain, perhaps reorganizing, cleaning-up, even changing the way circuits are arranged to operate more efficiently.[17] Others suggest that during sleep our minds play with images to try to work out how they go together and what they mean and where they may fit. This could explain why remembered dreams often involved unresolved issues—and why we only notice the dream itself if we wake in the middle of it. It could also explain why we each have slightly different perceptions of the same event. We each dream of different types of sheep due to our different reference memories: accountants dream of monetary sheep, artists of colorful sheep and rock climbers of mountain goats.

The fact that nerve cells can be altered by LTP, even grow and evolve opens another whole realm for us. It means that our brains and minds are much more flexible that most people have been led to believe. This flexibility remains for most of our lives, with researchers discovering that even people in their 60s and 70s are capable of generating new neurons. (More on this later). This could underpin evolution, albeit not quite as Charles Darwin proposed. It could also support the claim of some philosophers and many new age practitioners that conscious thought is the precursor of matter—as in electric-based thoughts creating news nerves.

While science has demonstrated that increased mental activity can grow nerves, it has also proven the opposite occurs. When neurons are inactive they can also shrink their connections.[18] In other words, use your mind or loose it. Interestingly, studies have found that just how much a person remembers can be predicted by looking at electrical activity in the brain's hippocampal region during learning.[19]

Other research has demonstrated that emotions are also associated with LTP. For example, researchers at McLean Hospital, in the northeast U.S. conditioned rats to fear a certain sound. They then examined neurons in the rats' brains and found that the sound, and fear, had generated LTP.

Through their work, they found that "an individual can have very poor conscious memory of a certain traumatic event, but at the same time, very strong unconscious emotional memories can be formed through a fear conditioning mechanism," said one of the hospital's researchers, Vadim Bolshakov. "These fears, which are very resistant to extinction, can become a source of intense anxiety."[20]

Extending this, other researchers found that such conditioned fear was able to be overcome by electrically stimulating parts of the brain. Researchers at the Ponce School of Medicine believe that stimulation to, and increased activity in the pre-frontal cortex strengthens feelings of safety and is able to inhibit the memory (or LTP) of fear.[21]

This research, and similar findings by others, takes the conditioning work of Pavlov and his salivating dogs to a whole new level. It provides a major revelation in that it helps explain how fear can be trapped in not only the conscious or unconscious mind, but also in the physical body—such as in nerves. It demonstrates that this is not just the result of a molecule, such as a peptide, becoming constrained somewhere, but that there is also a LTP resonance that is involved in altering nerves and energy flow, which in turn may also alter the flow of molecules as well.

Accordingly, in a situation opposite to fear, say one of euphoria, love or religiosity, could people associate a LTP resonance with what they might call the spiritual? It is more than likely. This could explain why some people have many spiritual experiences, they are conditioned or learn to experience them by the way their brains resonate and how they interpret this resonance.

Brain circuits and psychological experiences are not different things, rather they are different sides of the same thing, LeDoux suggests.[22]

In short, these discoveries suggest it is up to us as to how we interpret various LTP reverberations in our heads. If we interpret and label everything negatively, we will not only increasingly think that way, but also as feel and act that way. In contrast, if we consciously try to interpret, label and react positively regularly this should lead to more positive feelings and expression by our nervous system over time. Not doing anything leaves us somewhere in the middle and at the mercy of the world around us.

In terms of our investigation into our souls and spirituality, the creation of new nerve tissue by electric resonance, this creation of something out of nothing by energy, certainly sounds like a smaller-scale version of God's creation of the flesh of man as Michelangelo painted on the ceiling of the Sistine Chapel in The Vatican.

These electric circuits also get around the problem of atoms and cells being regularly replaced in the body. As if atoms in the cell or even the cells themselves are replaced by news ones, it is the flow of current and ultimately information that continues. This could also explain why memories change with age, as cells in the neural pathway are replaced and the electric patterns altered ever so slightly.

The next question appears to be is a spiritual experience a special LTP, or a stronger version of a LTP?

Possibly. However, it appears LTP is not alone as it is not peculiar to mankind, not something that sets us and our souls apart from the animals. Scientists have reported resonance in the nerve cells of giant squids, monkeys and other creatures. Though, the extent of LTP resonance does appear to be much greater in people than in animals. Human memory is known to comprise comparatively more subsets, such as explicit memory, which records facts and experiences; and implicit memory, which records conditioning, skills, priming and other memories.

This then raises another question, is the soul and spiritual experiences just another subset of learning and memory? Again, it is possible, but we do not know enough at present to determine this. Fortunately, while scientists research this further, it does not halt our investigation. There are still several other clues that we can examine.

Brain waves, the waves that result from the reverberation of nerve cells when we learn, think and remember, provide some interesting evidence.

Doctors have been measuring the electric waves in brains for decades with electroencephalograms, or EEGs. From these many measurements it is known that there are several categories of brain waves. These include:

1—Delta waves—occur at a frequency between 1–5 cycles a second or hertz. These slow waves generally only occur in adults during deep sleep. Interestingly, they sometimes appear in awake infants. (Is this the child-like state that we as adults are directed to return to by many religious and mystical texts?)

2—Theta waves—oscillate between 4–7 Hz. It is uncertain what role they play. Some research suggests it is at this wavelength or slightly higher (5–10Hz) that neurons resonate at for LTP. Other researchers claim this is the frequency at which we are most creative and insightful, while others claim these brain waves occur in children and adults experiencing emotional stress.

3—Alpha waves—resonate between 8–13 Hz. Virtually all of us emit these waves when we are awake and resting with eyes closed. When we fall asleep they slow into delta waves.

4—Beta waves—oscillate between 14–30 Hz and appear when the nervous and mental systems are active.

All waves are characterized by wavelength, amplitude, velocity and frequency. Wavelength is the distance from one wave crest to the next. Amplitude is the height of the wave. Velocity is how fast a crest is moving from a fixed point. Frequency is the rate of crests passing a given point (or wavelength divided by velocity) and is given as peaks, or cycles, a second (also called Hertz).

The power of a wave is determined by its amplitude; for instance the harder you pluck a guitar string the higher the amplitude, the energy and louder the tone. In contrast, less energy generally corresponds to a smaller amplitude and lower volume of sound.

Low frequency waves are relatively long and gentle, like swells in a calm ocean. Moving up the wave spectrum, wavelengths become progressively shorter and their energy levels increase. Ultra-violet waves, or rays as they are generally called, x-rays, gamma rays and cosmic rays are able to alter atoms and elements, changing them into charged particles with ionizing radiation.

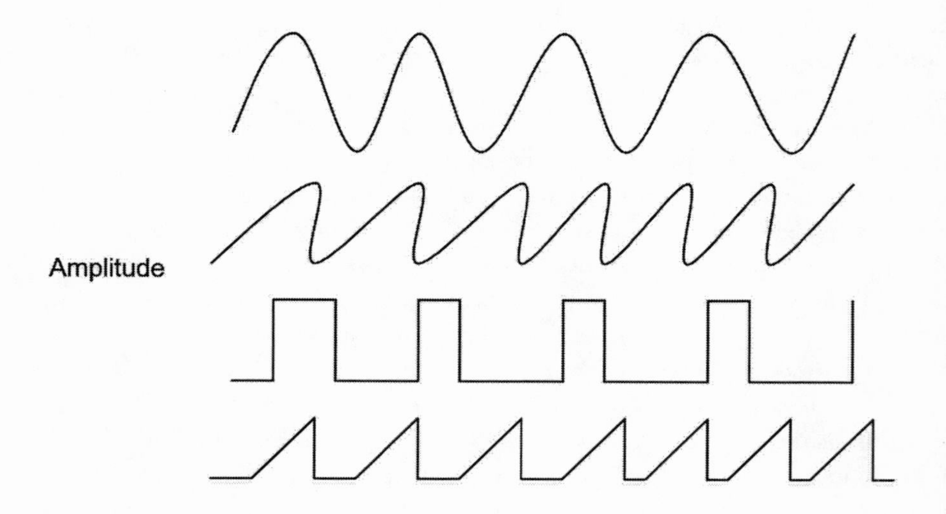

Figure 5. There are myriad wave types

With brain waves, it is believed that every thought has a pattern of energy characterized by particular wave pattern. For example, when you read the word "yes" here, we each produce identical brain waves, waves that can be accurately measured and identified. Yet, we each interpret this and other words slightly differently, even though we express similar brain waves.

Positive thoughts generally create a smoother, harmonious brain wave. For instance, people in trance states generally display smooth brain waves.[23] Rhythmic stimulation, be it controlled breathing, chanting, drumming, praying and swaying can also smooth the brain's electric patterns. Bob Beck, for instance, found a coherent brain wave frequency of 7.8Hz that is common to many clairvoyants and physics.

In contrast, negative thoughts invoke more irregular brain waves and EEG patterns. Interestingly, barbiturate-induced narcosis or altered states substantially increase brain waves, but only produce a minor change in the underlying electromagnetic potential of brain tissue. Conditions such as coma and epilepsy are also known to have distinctive brainwaves.

Brain waves do not resonate throughout the whole skull. For example, brain scans of meditating Tibetan Buddhist monks found brain waves diminished in an area that helps people orient themselves in three-dimensional space. Activity in the superior parietal lobe (toward the top and back of the brain) becomes inactive.

This creates a blurring of self-other relationships, says neurologist Andrew Newberg of the University of Pennsylvania. And if the inactivity lasts long enough the person may sense a dissolution of their precise location and instead sense a feeling of "spacelessness," a feeling of at one with the world and the universe. "When people have spiritual experiences they feel they become one with the universe and lose their sense of self. We think that may be because of what is happening in that [the parietal lobe] area—if you block that area you lose that boundary between the self and the rest of the world. In doing so you ultimately wind up in a universal state," says Newberg.[24]

Other research has found the prefrontal cortex (just a few inches behind the center of the forehead) is very active when people who are meditating focus intently upon something (as in active meditation or visualization), while other studies have identified increased electrical activity in other specific areas of the brain when people make morale decisions.[25]

These findings, and many more, show how different thoughts and mental actions involve different parts of the brain and different brain waves—and vice versa. Stimulating different parts of the brain can produce different actions and thoughts. Canadian doctor Wilder Penfield, for example, found that by applying a small electrical charge to the surface of an exposed brain he could trigger certain parts of the body to move. He identified and mapped the cerebral cortex, and which part triggered which body movement or created a certain perception or memory. By stimulating the temporal lobes (the lower portion on either side of the brain) and inside the cerebral cortex, he could prompt memories, colors and sounds, and often things a person had forgotten. Stimulating the same place prompted the same memory or reaction to arise, again and again. At the time, in the 1950s, this was proclaimed as a physical basis for consciousness.[26]

Spanish neurologist Jose Delgado has since expanded upon this and found that electrically stimulating other, generally deeper, parts of the brain produces not only mechanical action, but also emotions, feelings of pleasure or pain, euphoria or anxiety in people. Stimulating some of these areas produce major changes in a person's personality.

In one of his experiments probing this phenomenon, a group of rats were able to press a switch that provided food or another switch that provided pleasure (via an electrode implanted in a specific part of their brain). They invariably chose to press the lever that provided pleasure, to the extent that some of them died of starvation. This appears to be one of the few instances that a stimulant has been able to override the supposedly primary drive to survive.[27]

In another interesting experiment, Delgado was able to use extremely low frequency magnetic fields to induce sleep or manic behavior in monkeys. These currents were weaker than what was generally considered necessary to trigger a nerve cell.

In similar research, other scientists found that placing metal plates on each side of a person's head and sending a low frequency pulses through them not only alters a person's brain waves, but also there reaction time. Pulses between 8–10 Hz sped-up reaction times, while slower frequencies, such as 2–3 Hz, slowed reaction times. Taking this further, scientists claim they can provide the benefits of a good night's sleep to a person by delivering weak electric currents between the eyelids and behind the subjects' ears, currents that pulse at certain brain wave frequencies.[28]

Accordingly, the amount of electric current flowing through various parts of our heads in the form of brain waves obviously has a major role in not only mechanical physical actions, but also how we think and feel. But when we measure brain waves we are recording just one aspect of electric energy in our bodies. There can be many different types and shapes of resonating electric waves within any given period. Measuring and comparing these waves in terms of amplitude, frequency, shape, power (millivolts), distance (nanometers) and overall pattern is also important—and could provide for more useful information and analysis than the standard EEG. This range of descriptors of waves is very important, yet often overlooked by both by new age practitioners and scientists. This appears to be why some experiments and treatments work for some people and cannot be repeated for others.

Brain waves can obviously tell us a lot about what is happening inside our heads, but not everything.

When there are few brain or EEG waves, as when people are anaesthetized, it does not mean that we are unable to comprehend and remember things. It has been proven that the minds of patients under anesthesia can still recall what happens to their bodies during surgery, overturning the belief that this was impossible while their brain waves showed they were asleep. Accordingly, surgeons have to be very careful of what they say during operations. In fact, there have been reports of what surgeons say influencing patients' recovery, with positive statements speeding up chances of survival and recovery; and negative ones slowing it down or even being ultimately fatal. This shows the brain and mind continue to operate at some level even when brain waves appear dormant.

Resonating brain waves and LTP indicate the power of an immaterial energy. Could a similar power be involved in how our souls experience and express spirituality?

5—A Body of Power

Besides the waves in our brain, there are other electric waves being generated and resonating in our bodies. Organs such as our hearts also produce significant electric waves.

The waves of a heartbeat are triggered by an electric pulse in the muscle around the heart. Medical textbooks attribute the beating of our hearts to rhythmic electrical activity, with the source being autorhythmic cells in the heart muscle.[1] It is these cells that allow a heart that has been removed from a body for transplant to eerily continue to beat on its own, even though all of its nerves and other connections have been severed. This appears to be a variation on LTP, a resonance of cells designed to keep the heart beating without external input.

Like brain waves, heart waves can also be measured, and modified. The electrocardiogram, or ECG, which measures heart waves is a standard tool for medical diagnosis, much like the EEG is for brain waves. The magnetic field generated by the electric currents of the heart (about one-millionth of the magnetic field of the earth) is stronger than that created by the electric currents in the brain.

Some scientists suggest the heart's network of nerves is similar to those found in the cerebral brain, and that it is in fact a "little brain".[2]

Nerve impulses from the autonomic nervous system can make the heart beat faster or slower. When we sense danger nerves in the autonomic nervous system send signals to the heart that prompt it to beat faster to help us either "fight or make flight", whether we face and fight a potential danger or run away from it. This also triggers a subsequent cascade of chemical signaling within our bodies, such as adrenaline and so on, that facilitates the ensuing mechanical action.

During moments of frustration, stress or insecurity, the waves are erratic and jagged—and interfere with each other, and other waves in our nervous system. This appears to be one reason why we can't think clearly when we are angry or stressed.

In contrast, when we experience pleasant emotions, changes in the electric waves of our brain and heart can be rhythmic and smooth enough to dance to.

Besides the heart, other body organs also emit electric waves. Some preliminary research suggests this may be the case with the liver and it is known that the stomach has millions of nerves that emit electric waves.[3] This gives some credence to the adage "gut feeling". It could also explain why we sometimes feel butterflies in our stomachs, as the nerves resonate and tingle under certain conditions, causing this sensation.

It is also known that bone generates a minute amount of electricity when compressed or stressed, called pizoelectricity.[4] This is a a mechanically-induced energy and quite different from the ionic-cellular and resonating LTP electric energy.

Other elements of the body are conductors of electric waves. The circulatory system is a good conductor due to the salt, iron and other elements in blood; allowing ECGs to be taken from almost any part of the body. And when tissues with different electronic properties meet in the salty fluid of the body, a sort of 'diode' is formed.

Aside from emitting electric waves, some organs are also able to sense energy. Every human sense, for example, is able to detect nuances in energy and send what it senses, as electric signals, via the nervous system to the brain and body to determine how to respond.

Sight is the response to electromagnetic radiation of wavelengths between 400 to 700 millimicrons, focused through our corneas onto the retina. Hearing is the response to sound waves impinging on the eardrum, the hammer, anvil and stirrup bones, which send waves into the cochlear, where the corti differentiates them and sends them via nerves to the brain.[5] Smell is sensed by olfactory receptors, where the actual odor binds to the sensing cell membrane and again triggers one or more nerve impulses to the brain.[6] Our skin responds differently to various wavelengths or energy, for example getting sun burnt by some frequencies.[7]

Interestingly, there is also an emotional aspect or tag to many of these senses, as we often experience with taste, touch and the like. For example, a sweet taste often promotes a positive emotion, as can a hug. Besides the sensation creating an emotional response inside us, an emotional response can also impact the senses. For instance, emotions can change the electrical potential of our skins, as demonstrated by lie detectors, which measure the psychogalvanic response, a momentary decrease in the electrical resistance of the skin.

As can be seen from all this, the whole human body uses and is sensitive to energy in a wide range of forms. We just don't notice it in our day-to-day lives.

What is the energy and the sensation process involved in spiritual experiences? How does a mind become aware of itself and consciously comprehend things spiritual and its place in the scheme of things? To answer this, we first have to understand the consciousness.

6—Consciousness Considered

Consciousness is interesting in that while we are all individually experience consciousness, we know comparatively little about it in scientific terms. It is more than what we loose when we go to sleep and get back in the morning when we wake up.

What we do know is that circuits of nerves in our bodies and minds labor ceaselessly, processing enormous amounts of information—and most of the time we are not even aware of it. You duck, you jump out of the way of a ball or make other snap decisions with the information reaching your consciousness split sections later. Routine tasks are not consciously recognized, not even remembered. Did you turn off the stove, lock the door today? In fact, it is estimated that 90% of the brain's workings are at the unconscious level. A large part of your mind's day is spent in an unconscious state.

This unconscious nervous system keeps the body functioning—and stops the conscious mind from being overloaded.

If we are unaware of these activities what else are we unaware of in our daily lives? Lots.

Studies show the brain unconsciouslly recognizes subliminally presented information, such as identifying words that a person is not conscious that they have seen—such as the next sentence.[1] You just aren't consciously aware of it, yet. Now you are.

When you talk, you generally aren't consciously thinking ahead about every word you are going to say, but your subconscious is. Similarly, huge chunks of your brain, far more mass than what seems to be involved in higher reasoning skills such as language, have evolved to unconsciously and subconsciously work out problems such as turning a page and comprehending printed words.

Most of our day, we are hardly aware of all of the things our brain broods upon. How often, for example, have you felt moody or glum without knowing the reason? That's your subconscious working, at the level just below consciousness, but not at a level that you are fully aware of.

Only a comparatively few elements of unconscious thought get through to consciousness. Maybe this is why the word "consciousness" is both singular and plural. It should be noted, that while they are related, the mind does not equate to consciousness.

How can we look at words on a page, say this page, but not understand the lines and curves; not until we concentrate upon them and recognize and reconcile them with what we already know, what we already have filed in our nerve circuits. If you do not read English you might be conscious (as in terms of being aware), but you would not be able to be conscious of the meaning of these lines and curves. The more words you understand the more conscious you are of their meaning.

Neurologist and author Antonio Damasio suggests consciousness is the result of the nervous system's unconscious dialogue between the brain and the body, (such as the stomach). Accordingly, the whole body as well as the brain plays a role in the perceiving and reasoning process. The process of maintaining internal stability, or homeostasis, by sensing the world and changes around us and adjusting ourselves to them creates an intense communication that sometimes overlaps itself and ultimately becomes conscious. In his somatic marker hypothesis, Damasio suggests emotions are also a result of communications between the brain and the body. As such emotions are mixtures of brain states and bodily experiences. This gives credence to the role of "gut decisions".[2] Other research has since found that the stomach contains a nervous system capable of learning and memory as well as communicating with the brain. Every neurotransmitter that exists in our brain also exists in our stomachs.[3]

In contrast, other scientists say consciousness is simply the result of the mechanical firing of nerve cells and that our sense of conscious self depends to some degree on interconnected neurons in the prefrontal cortex of the brain.[4]

Nobel Laureate Francis Crick suggests consciousness specifically depends on connections between the thalamus and the cortex and that it exists only if certain areas of the cortex have "reverberatory circuits". He theorizes consciousness consists of the synchronized firing of neurons in the range of 40Hz, and even suggests the location for this is between the thalamus and layers four and six of the cortex of the brain.

Crick also says that while the brain and nervous system is in place at birth, it grows and is tuned by experience and learning. Experience tunes the hardware until it can perform in a more precision manner. Citing instances of patients with brain injuries, he also suggests the anterior cingulate sulcus is the site in the brain where free will originates.[5]

Other scientists believe consciousness is an electrical threshold that nerves must fire above to bring an experience into consciousness. You might be aware of something, but you can't bring it into consciousness because you haven't focused enough resonating nerve cells on it yet. "The hypothesis is that this oscillation [of waves] is the medium through which the neurons act together by becoming synchronous," suggests French researcher Fransisco Varela. "They tune their oscillations together. It is like a transitory glue. These transient patterns of synchrony are certainly something that relates closely to the moment of consciousness."[6]

Similarly, John Hart suggests conscious memory is the result of the uniting together of electric waves in different brain regions. Hart and his team suggests "memory appears to be a constructive process in combining the features of the items to be remembered rather than simply remembering each object as a whole form". They found the thalamus regulates rhythms that connect brain regions and the various components of memory. "The thalamus seems to direct or modulate the brain's activity so that the regions need for memory are connected. It appears that the electrical signals synchronize the brain regions that store each part of an object's memory so that those areas are connected. This co-activitation of brain regions likely represents the memory of the object itself. It may also explain why we may remember something clearly, and other times we can only come up with parts of the item we are trying to remember.…This may occur when the rhythms don't synchronize with the regions properly. It could also explain why the memory will come to you at a later time," he explains.[7]

Nobel laureate Gerald Edelman suggests consciousness is a chain of events, with rapid back-and-forth signaling between various groups of nerves. Mutual messaging between neurons, such as those involved in perceiving and those involved in memory and responding, create a unity of consciousness. He describes consciousness as both "integrated" and differentiated," with the former being when neurons interact with themselves and neighbors, while differentiation involves their acting with other nerves elsewhere. The more differentiated groups of neurons are, the more they interact rapidly, the greater their contribution to consciousness.

Other studies have found that the neural patterns in the brain differ when a subject is viewing an image subconsciously and when they are paying attention and viewing it consciously. Neurons fire in different patterns depending on whether the subject is viewing the image or merely recollecting it, demonstrating the importance and influence of nerve circuits to different levels of consciousness.

Susan Greenfield proposes that the firing of neurons form what she calls "transient assemblies," or short-term networks, interact with other nerve signals in the brain and body and elevate their messages to consciousness. The degree or depth of consciousness correlates with the size of the transient neural assemblies.

Accordingly, consciousness is not an on or off property, but rather a sliding scale, continuum that can grow and evolve. Interestingly, human newborn babies have the least mature brains of any mammal at birth, with only 25% of adult brain size.

If consciousness involves a sliding scale or series of degrees, this could explain how some people who are unconscious, either sleeping, anesthetized or hypnotized, can recall some things afterwards while others are not—people have differences in their sliding scale.

Greenfield suggests the more the mind predominates over raw emotion, the deeper the consciousness. (If this is the case, it suggests the purpose of life, individually and collectively, is to increasingly become more aware and conscious.) She notes how the greater the force behind the stone thrown into a pond the greater the ripple that results, and says this analogous to consciousness, with the greater the neural assembly the greater the depth of consciousness. In *The Private Life of the Brain* she suggests there are several factors involved in consciousness:

1—the number of neuron connections. The greater the assembly the great deeper the consciousness;

2—the appropriate chemical environment to facilitate the nerve connections;

3—the degree or force of stimulation; and

4—the speed of assembly formation.

Another theory of the physics of consciousness has been developed by English mathematician Roger Penrose and American psychology professor Stuart Hameroff. Their Orchestrated Object Reduction theory suggests that inside each nerve cell, microtubules of protein can act either as particles or waves.

These microtubules dance with thousands of others in a coherent network. When stimulated enough, they collapse from their wave/particle duality into a singular, classical process that fires neurons and forms conscious thoughts. They note the tiny microtubule of a neuron cell has a slot on the side within which an electron resides. Making someone unconscious immobilizes the electron and the microtubule—and electric signals underlying normal electrochemical exchange. The repetition of these changes, like LTP, at high speeds and great quantity could develop a ribbon of continuity of consciousness.[10]

LeDoux proposes that what we are conscious of are the perceptions that working memory is working on. He also believes that language embellishes working memory, and in doing so makes human consciousness unique.[8]

Going back to the example of words on a page, you can only be conscious of their meaning once you have knowledge and/or experience of what the particular words mean—a conditioned perception and reference that matches a particular electric circuit, and vice versa. The same applies to language, with symbols and sounds having much greater meaning than their physical base.

It would appear that the electric currents, or "electric language" flowing through our nerves, does something similar, with each pattern of impulses carrying information and meaning. Currents are sent from senses in encoded patterns and decoded to produce perceptions, meaning and memories.

Accordingly, these currents are like the words on this page, they can and do have particular meaning to you. A current in particular nerve circuit could similarly represent a particular perception, just like a word can be descriptive. A current in other circuits could be more adjective or verb-like, recalling emotions and feelings or actions. Sometimes these circuits come together as words in a sentence, with different currents flowing in different circuits to provide greater meaning and consciousness, even those varying degrees of consciousness.

Another element would be the ability of memory to recall and retain those words, or circuits, in an "electric sentence". This is like the way that various musical notes, which make little sense on their own, can flow into one another and create a piece of music.

In the case of nerve circuits and "electric words", this requires coherence of electric currents. The more coherent the electric waves, the more coherent the referencing and meaning can become among those circuits. In theory, the greater the coherence of electric waves, the greater the level of awareness and consciousness.

This process is a little like radio waves, with the senses of sight, touch and so on being the transmitters and the brain, heart, stomach and so on being the receivers. (Emotions may be added as further encoding, or as booster transmitters, or maybe a mix of both.) The brain and other receivers decode the messages that we can't see flowing through our bodies, much like we can't see radio, television, wireless and other waves flowing through the air! When the information is flowing, we are not conscious of it. We are only conscious of it when it is properly decoded and put into context.

The late English physicist David Bohm suggested this electric wave of electrons was a form of active information that was both physical and mental in nature—and formed a 'bridge' between the two. Information contained in thought, which we feel to be on the 'mental' side, is at the same time a related neurophysical, chemical and physical activity, he suggested.

This is where the physical quantum world intermingles with the mental world of consciousness.

Then there are also levels of consciousness that we only rarely glimpse—and are still trying to understand and attribute meaning. These are often considered spiritual. But just how is our consciousness raised above the everyday, tuned to an extreme on that sliding scale?

Consciousness appears to involve an intriguing property of waves, a property called stochastic resonance. Stochastic resonance is where a stochastic fluctuation or wave of background "noise" joins together with other weak (often undetectable) waves to make them stronger. The noise wave adds to the other wave, amplifying them and increasing them to a threshold where they can be detected. In short, this small force, like pushing a swing that is already in motion, can eventually have a large effect.[11]

Scientists have found that this principal increases the ability of nerves to recognize previously undetectable signals.[12]

This stochastic resonance can also play a part in getting nerves to continue to fire and resonate as in LTP, with different nerve circuit waves joining together to keep neurons firing over extended periods of time. In neurons, stochastic resonance is known to involve the transfer of ions between cells, in particular calcium signalling, lipid membranes and microvillar F-action bundles, with the latter functioning like little electronic devices that can cohere cation transfer.[13]

Stochastic resonance is also known to be able to enhance other human senses, such as touch and balance.[14]

And it appears that stochastic resonance can also heighten our consciousness.[15]

With this stochastic process amplifying previously undetectable waves, or sensations, we also have a scientific understanding of "extra sensory perception". Stochastic resonance provides a way to heighten peoples' senses and awareness.

Can this property, when coupled with coherent resonance of brain, heart and other waves, also lead to spiritual experiences? Are spiritual experiences then just an elevated form of consciousness, a result of our conscience? Possibly. The soul is often said to amplify the elevated elements of our minds. Or is a spiritual experience a "peak" mental experience, with nerve cells resonating around one perception? Or is it more the opposite, a coherent resonance that arises when we don't think about anything inparticular and just let our mind flow? Or is it like any other unconscious activity, with only limited realizations being brought into consciousness?

What is known for certain is that while a brain looks the same whether it is conscious or unconscious, the electrical patterns inside it look very different. During every conscious moment, the brain is abuzz with electrical activity and can been imagined as a magical sphere in which lights incessantly twinkle as myriad nerve cells fire on and off, with some resonating in varying degrees of unison, peaks and valleys. In fact, the electrical pattern of the whole nervous system is quite different when we are conscious compared to when we are not.

It appears that the nervous system is a little like a car battery. While the molecules in the lead battery can generate electricity to power lights, start a car and do other amazing things, they sit idly until the circuit is completed with the ignition key. In terms of consciousness, the sensing of something and that signal reaching the appropriate nerve cell/s is what completes the circuit inside us and then powers the perception. potentiation and response.

Unfortunately, being conscious can help us think and reason, it does not necessarily make us spiritual. Animals are obviously conscious, but not necessarily spiritual.

To be spiritual, many religious and spiritual texts say we need the highest levels of consciousness, conscience rather than just consciousness. Conscience is often regarded as a moral compass, something inside us that helps tell right from wrong.

But how can impartial energy, atoms and nerves do this? What makes cells potentiate? Do they just do it on their own volition, or is there a little 'brain' in them telling them what to do, or is it something else?

Besides noting that unconscious thoughts precede conscious ones, Benjamin Libet examined whether the brain approves or vetos unconscious impulses when they are made conscious. He found that conscious intention followed an unconscious or "preconscious" process and permitted or inhibited it.[16] (The learning and memories created and recalled by LTP are used as references in making these decisions.)

Libert's research demonstrated that while you may not be able to control what your unconscious is working on, you can control whether or not to act on it when it becomes a conscious thought.

This choice to act or not is what provides us with self control. This overriding power is in accord with many religious and social views of ethical behavior, individual responsibility—and the functioning of your soul.

Consider the response of two people who notice $100 lying on the pavement outside a bank. One person might swoop down pick up the money and put it in their pocket and quickly walk on. Another might pick up the money and take it into the bank and suggest to the manager to that fact that someone may have dropped it. How each person perceives the enviroment and how they reference the situation, their past experiences as recorded by their nervous system and molecules of emotion, is what determines how they consciously respond.

Delgado suggests, "The uniqueness of voluntary behavior lies in its initial dependence on the integration of a vast number of personal past experiences and present receptions."[17]

LeDoux points out the mind is not just a thinking device. Rather it is an integrated system of synaptic networks that process cognitive, emotional, and motivational functions.

He adds, "We all have the same brain systems, and the number of neurons in each brain system is more or less the same in each of us as well. However, the particular way those neurons are connected is distinct, and that uniqueness, in short, is what makes us who we are. Your 'self,' the essence of who you are, reflects patterns of interconnectivity between neurons in your brain.

"People don't come preassembled, but are glued together by life. And each time one of us is constructed, a different result occurs. One reason for this is that we all start out with different sets of genes; another is that we have different experiences. What's interesting about this formulation is not that nature and nurture both contribute to who we are, but that they actually speak the same language. They both ultimately achieve their mental and behavioral effects by shaping the synaptic organization of the brain. The particular patterns of synaptic connections in an individual's brain, and the information encoded by these connections, are the keys to who that person is."

In short, our minds, our personality and conscience are shaped by the genes we are born with and molded by the learning and experiences we have throughout our lives.

LeDoux reminds us that as the brain is modified by experience, so too must synapses be changed by experience.[18]

Yet, while the synapses and molecules of our body change in this metamorphosis, others recognize that we are the same individual. Interestingly, most people rarely consciously try to develop some sort of personal identity. Rather, we journey through life unconsciously evolving our personality. Maybe we should take a more conscious and conscientious approach to developing our selves, our souls.

7—Another Powerful System

There is a second electric system in our bodies, one that is less known than the nervous system taught in schools, but none-the-less very important to us, and our search.

While nerve cells convey an on-off current like that in a computer, there is also a direct electric current that flows continuously in our bodies. Underlying the rhythms of our nerve cells are direct electric potentials that vary much more slowly, over periods as long as several minutes, notes American biologist Robert Becker and one of the early discovers of this bioelectric system.[1]

Becker identified positive electric potentials over each lobe of the brain, the spinal cord, the brachial plexus between the shoulder blades and the bones at the base of the spinal cord. In contrast, he found hands and feet have negative potentials. Very weak electric current flows continuously between these potentials.

At first glance this suggests electric current flows from major nerve centers to the extremities of our bodies, from positive to negative.

However, possibly even more importantly, Becker also identified a difference in the electrical current of nervous systems, in particular motor and sensory nerves, with one positive and one negative, allowing for a complete electric circuit in our bodies.

He found that these two types of nerves, motor and sensory were polarized in opposite directions. The voltage of one system was positive towards the toes, while the other was polarized in the opposite towards the trunk, and always had a higher voltage gradient. Becker found a 4 millivolt positive current in one branch and an 8 millivolt negative current in the other branch over a one centimeter length of frog leg. This created a loop or electric circuit.[2]

This shows that rather than one electric flow from head and spine to fingers and toes there are complete electric circuits flowing around different nerves and that there are different voltages in different nerves.

These continuous currents are much weaker and slower than the on-off digital currents of our nervous system, they pulse in strength between 2–30 cycles a minute, that is a minute not a second. They are also very weak and are measured in microvolts.

Generally, these perineural currents are not noticed in brain or other scans, being filtered out during an EEG for example. Yet they might be as important, or even more important, than the more familiar brain and heart waves.

Becker and other researchers found this continuous current is conveyed by perineural cells, in particular glial and Schwann cells, which encase our nerve cells.

It appears that every nerve in our bodies is encased by perineural cells, with these types of cells found wherever there are nerves. (When Becker grew perineural cells in culture, he found an occasional nerve cell also grew.) And while there are billions of nerve cells in the brain, there are even more perineural cells, with glial cells encompassing nerves in the brain and spinal cord and Schwann cells wrapping around nerves in the peripheral nerves. The interaction of nerve and perineural cells appears to be analogous to how our pulse is to breathing: both are important in the exchange of energy in our lives. Faster breathing generates a faster pulse or heart rate, and vice versa.[3]

They also appear to have some properties of electric circuits as found in the inductors, capacitors, transistors, gates and other elements that are used in integrated circuits of computers. For example, when a capacitor and an inductor are connected in parallel the capacitor discharges through the inductor, creating a magnetic field and when the field collapses it produces a current that charges the capacitor again (in a process akin to ion exchange in cells). When this process is rapidly repeated at a rate that best matches the capacitor and inductor a resonant frequency is generated. When an alternating current is added that matches this resonant frequency the circuitry produces a larger output than the individual currents alone. Is this what happens in perineural cells?

While the role of perineural cells is not well understood at present, it appears there is an element of electromagnetic current interaction and regulation between them and the nerve cells they encase. Becker found the continuous current in the perineural conveyed by electrons. (Are these the electrons referred to by Penrose and Hameroff's theory of consciousness?) In contrast to ionic energy flows, which tend to weaken after short distances, electrons can travel much longer distances. This continuous flow of electrons provides not only an electric current, but also generates a surrounding magnetic field, which can in turn affect other electrons nearby.

The overlay of one electric circuit upon another can influence both, with parallel electric currents able to generate attraction and repulsion between them. Interaction between nerve and perinueral currents appears to exert some control over the other, either speeding it the current or slowing down interactions.

Other research into these perineural currents suggests they also play a role in getting those chattering pyrimidal neurons and other nerve cells to resonate and LTP so that we can learn and remember. It appears they regulate the nervous system to some degree, probably by creating localized electromagnetic fields around the nerve cells which could help strengthen and cohere the LTP resonance.

"It appears that the DC [direct current] is somehow involved in getting the neurons ready to fire the command to move the muscles," notes Becker. The electromagnetic field can help ready a neuron to send a signal, so that a small stimulus can cause it to fire, compared to a larger signal being required if there is no support from the local field. He says, "This phenomenon, which has become know as the 'readiness potential,' seems to imply that the DC system commands the nerve impulse system."[4] Other researchers have since found that it appears the readiness potential precedes the decision to move a muscle.[5]

Overall, Becker discovered these perineural direct currents within the central nervous system regulated the level of sensitivity of neurons by several methods:

- by changing the amount of current in one direction,
- by changing the direction of the current (reversing the polarity), and
- by modulating the current with slow waves.

These features are quite different to the on-off digital activity of nerves and creates some significant additional benefits for life. Becker notes, the direct current nervous system seems directly involved in every phase of mental activity. The electric sheath of perineural cells act as a sort of background stabilizer that keeps the nerve impulses flowing in the proper direction and regulates their speed and frequency.

He suggests the perineural cells and current also probably play an active role in the mind. Variations in the current from one place to another in the perineural system apparently form part of every decision, every interpretation, every command, every vacillation, every feeling, and every word of interior monologue, conscious or unconscious, that we conduct in our heads, he says.[6]

Could a spiritual experience be a joint nerve and perineural LTP? Only further research will tell.

The fact there are two electric nervous systems working in our body is a major discovery. Two information and_sensing systems, sharing and exchanging information—and where one can influence the other.

While the interaction between these two systems is comparatively little understood at present, with research into the field just beginning, there are some significant findings that suggest we are on the right path here in terms of our investigation, and which could help explain some aspects of the soul and spirituality.

For instance, this electromagnetic perineural system is known to be involved in growth, healing and other activities that have often been ascribed to things spiritual.

Becker, for example, found that when bones heal they emit a positive electric charge, while the periosteum nerves around them become sharply negative and remain so while they heal. The polarity difference is wide soon after injury, gradually diminishing as the injury heals. He noticed this difference created an electric field that aligned collagen molecules among others to aid healing. Not only were there two perineural currents, they also produced electric potentials of opposite polarity, acting like the electrodes of a battery. These living electrodes were creating a complex field whose exact shape and strength reflected the position of the bone pieces. He identified similar current flows in the healing of wounded salamanders and found the polarity of the current reversed right after an injury, such as an amputation, from negative to positive and climbed to 20 millivolts positive. Then between six and 10 days, the polarity changed again, rising to 30 millivolts with a negative polarity as the salamanders began to regrow their limbs before returning to a baseline of 10 millivolts negative at their limb extremities. He also tested frogs, which were unable to regrow amputated limbs—and found no change in current polarity (unlike in the salamanders).

This current was also involved in the regrowth of amputated limbs of rats. "Small amounts of negative electrical current administered to an amputation between the shoulder and elbow turned on a regenerative process," he notes. "An actual blastema [of healing cells] formed and grew into all of the missing structures, including bone, muscle, nerves, tendons, and so on to the elbow joint itself. While we could not get the growth to proceed beyond this point, the experiment nevertheless proved two important points. First, electricity was clearly a stimulus to regeneration. Second, and more importantly, the instructions required to make a new leg were still retained by mammals."[7]

Using simpler organisms, biologists March and Beam were able to grow a head on the end of a flatworm where a tail would normally grow—by applying a small electric charge and reversing the polarity.[8]

In another important finding, Becker found an electromagnetic perineural current between nerves and the epidermis to be involved in the healing of skin.

This electric current creates a field that instructs cells to "dedifferentiate" into other cell that could help heal the wound! Becker noted that nerve fibers joined the epidermal cells like plugs into sockets to complete the circuit that provided the current for dedifferentiation. The neuroepidermal junction over the end of the damaged stump continually produced blastermal cells exactly where they were needed, at the healing tip. This important discovery "proved beyond doubt that the electrical current was the primary stimulus that began the regenerative process, and that it could operate even in mammals," says Becker.[9]

More recently, Catherine Verfaille identified what she dubbed "multi potent adult progenitor cells," which are able to form dedifferentiated cells from a wide range of tissues and create other cells.[10]

Other studies have identified that applying DC electric fields enhances the out-growth of nerve axons, toward cathodes, with the axons aligned along the lines of current flow.

In another fascinating finding, researchers in the late 1990s found electromagnetic force could directly affect our genes, our very DNA. Betty Siskin and her team demonstrated that electromagnetic fields could turn on genes in damaged nerves, genes that then played a role in triggering growth-related repair.[11]

These discoveries that show that electricity is involved in healing and regrowth, and also prompts once-specialized cells to transform into other cells, is a major scientific advancement.

This cell dedifferentiation is also now known to even apply to red blood cells, where the nuclei is virtually inactive but is able to be reactivated by an appropriate electric current. In an important point, Becker notes, "for electricity to produce an effect upon cells, there must be not only the right amount of electricity of the right polarity, but also cells that are sensitive to DC electrical currents." Becker suggests the best current for dedifferentation of red blood cells ranges between 200 to 700 picoamps.

In an interesting aside, he found that pulsed (instead of continuous) electric fields did not prompt regrowth, but rather only had an effect on cells already healing. "The mechanism of growth stimulation by pulsed magnetic fields was a magnetic effect on cells in active mitosis, not a turning on of the normal DC growth-control system."[12] This could explain why magnetic therapies work for some people not others; the correct magnetic field could aid healing already in progress, but not turn it on, or necessarily turn pain off. Subsequent research has also shown that a pulsed electromagnetic field can increase the growth of cancer cells in culture.

This electromagnetic perineural current also displays several other properties and functions, functions that have been interpreted as pertaining to things spiritual.

Just as he found he could turn growth on with negative electricity, he also found he could turn it off with positive current, stopping regrowth and healing—and even consciousness, Becker found. He measured the currents in animals under anesthesia and found that as each animal became unconscious, its peripheral voltages dropped to zero, and in very deep anesthesia they reversed to some extent with the limbs and tails going positive. They reverted to normal just before the animal woke up. He discovered that if he reversed the current flowing through the head of a salamander, he could cancel out its brain waves and make it fall unconscious, become motionless and unresponsive to pain. However, a chemically drugged salamander could only be partially awoken by applying the current of a conscious animal.

Similarly, Delgado found that he also could change the consciousness of monkeys, even put them to sleep by electric stimulation of certain parts of their brain. Such stimulation has also been found to be able to stop bulls charging and alter a range of other emotional behavior in monkeys.[13]

Not only can these electric currents be used to turn consciousness on or off, they can also be used to vary the level of consciousness. Becker and his team found that in hypnotized people the negative potential of the front of the head became less negative, often reaching zero, as the subject slipped into a trance. The electrical reading changed in the same direction as in anesthesia, only not as far, he found. "Some doubters (including myself, I'm afraid) had believed hypnosis-analgesia was merely a state in which the patient still felt the pain but didn't respond to it, but these experiments proved that it was a real blockage of pain perception," Becker says. "It seems that the brain can shut off pain by altering the direct-current potentials in the rest of the body 'at will'."[14]

He also noted "initial spikes," sharp sparks of electric signals that appeared to start or signal a change in brain waves and consciousness. These spikes appear to be crucial to getting the brain and body's electrical network to know what is coming and prepare it for action.

Also, Becker discovered he could alter consciousness by placing a very strong magnet at right angles to divert and decrease the current within salamanders' brains to the point where they would become unconscious, as if they were anesthetized. This required a powerful magnet (with 3,000 gauss) to divert the current within the salamanders' heads.

Other researchers have since found that both the immune and circulatory system can be altered by electromagnetism.[15]

In another study, physiologist Valerie Hunt took electromyographic recordings of a dancer as she entered an altered state of consciousness and found that the electromagnetic signals from the muscles in the woman's arms and back decreased and stopped, while electromagnetic energy poured out of her head while she was in the altered state.[16]

Hunt refers to another experiment she undertook in a room where the amount of electromagnetism can be altered, located at the University of California Los Angeles' Department of Physics. When the electrical aspects of the room were reduced, subjects were uncertain as to the exact location of their bodies. In contrast, when the electric field was raised above normal, the subjects' reported an expansion of their consciousness.

When the magnetism in the room was lowered, while the electrical aspect of the room remained normal, the subjects could not balance and had difficulty performing simple coordinated moves. In contrast, when the magnetic field was increased above normal, their coordination improved and they could balance themselves more easily than normal.

Becker also suggests emotions can impact the electromagnetic perineural currents through a an intricate and multilayered self-regulating feedback arrangement. He suggests that while a person's emotions affect the efficiency of healing and the level of pain on a psychological level, there is also "every reason to believe that emotions, on the physiological level, have their effect by modulating the current that directly controls pain and healing".[17]

Anatomy professor Harold Saxton Burr claimed he could predict a person's physical and emotional state due to changes in the voltage between their head and their hand.

In another experiment by Burr, a comparison of psychologically disturbed mental patients to normal subjects "showed clearly enough that the group consisting of those markedly deviated from normal behavior by psychiatric examination also showed a deviation in electro-metric examination".

Burr took this further and theorized "there is a very close relationship between the genetic constitution and the electrical pattern. If further studies should confirm this conclusion, it seems very probable that one of the ways the chromosomes impart design to protoplasm is through the medium of an electro-dynamic field."

This discovery that electromagnetic current is associated with the level of human consciousness is an important finding for our investigation into the soul and spirituality. Sometimes people report spiritual experiences as equating with a change or elevation in consciousness, with countless new age writings attesting to this. Numerous scientists, psychologists and philosophers have also questioned if spiritual experiences are a type of altered states of consciousness.

If this is the case, we have identified a mechanism this is key in varying levels of consciousness.

Becker himself suggests the DNA-RNA apparatus isn't the whole secret of life, but a sort of computer program by which the real secret, the electrical control system expresses its pattern in terms of living cells. This pattern is part of what many people mean by the soul, he says.[18]

This role of electromagnetism in consciousness, growth and healing as well as directing cellular and genetic activity sounds a lot like the "spark of life," "energetic breath," "life energy" or "creation force" referred to in antiquity and religious descriptions pertaining to our soul and spirituality. Is this just semantics or reality? You be the judge.

The only problem is that there are similar current in plants—which obviously do not appear to have a soul or spiritual experience. For instance, Russian biophysicist A. M. Sinyukhin[19] identified direct current electric currents in tomato plants in the 1950s. He cut a branch from plants and noted that a negative current of electrons streamed out from the cut for the first few days. Then as a new branch began to form, the current became stronger than reversed its flow to positive. The positive current increased the metabolic rate of cells in the area, cells that also became more acidic. To see what happened if he tried to assist the regeneration current, Sinyukhin applied a further current of just a few microamperes (a millionth of an amp) and found the plants restored their branches three times faster than before. If he applied a current of the opposite polarity it stopped the growth.

Burr found similar electric fields for trees, which surged during the full moon. They also reacted to storms, sunspots and other geomagnetic events. "Statistically, both the sun and the moon seem to influence tree potentials, the sun apparently through an electro-magnetic mechanism, the moon through a gravitational or gravitational-electrical mechanism," suggested Burr.[20] It appears that trees just have this one type of continuous electric current, and that the voltages change as the seasons change.

Becker also found changes within electric potentials of animals as seasons changed, even though the temperature remained the same in his laboratory. He theorizes animals' continuous electric system is influenced by the energy fields of the earth, in particular the earth's magnetic field. This is interesting, as the earth's magnetic field is very weak, about half a gauss and was previously thought not to be an influence, especially considering the other magnetic forces around us in the modern world. However, Becker notes that this is accompanied by micropulsations ranging from less than one to about 25 cycles a second, the latter being similar to the frequency of brain waves in a conscious person.

Ultimately, Burr suggested all living things are formed and controlled by electromagnetic fields. He believed that these "fields of life," or "L-fields" as he called them, were the basic blueprints for the growth and life, and expounded upon this theory in his book *Blueprint for Immortality: the Electric Patterns of Life.*[21]

Burr found electromagnetism to be involved in the initial creation of human life. He discovered changes in the electric potential of women during ovulation and suggested that the initial rise in this energy field was not only related to, but also brought on, ovulation—the process responsible for the creation of human life. He noted these changes could be identified by changes in the electric field between fingers of each hand. The electric potential before ovulation is about 50 millivolts negative, until about five days before ovulation when it begins to rise and becomes positive a few days before, rising to around 75 millivolts positive. But after ovulation it drops rapidly to zero then gradually decreases over the following 10 or so days back to where it started. Burr stated ovulation was one of the body's many oscillating electric fields and noted it was difficult to filter and isolate it from the many others, but that it could be done (and used his recording process to develop a fertility control system for which he was granted a U.S. patent).

In an interesting aside, the substance that stimulates the flow of breast milk in nursing mothers, the pituitary hormone prolactin, seems to sensitize cells to electric current.[22] Scientists have also identified a "calcium spark" or wave that surges through the human egg after fertilization and prompts it to divide and begin forming an embryo.

It appears the perineural current is much more developed in humans, than in trees and animals, and when combined with the complexity of the human nervous system, we have substantial differences from the nervous and electromagnetic current systems of vegetables and animals.

For example, it appears that both the digital on-off nervous and continuous perineural, currents are required for human consciousness. Having the two working together provides greater power and effects than either on their own or even the sum of the two added together.

Becker, for instance, found that by adding an extremely low frequency (ELF) on-off current of between 1–10 Hz to a continuous perineural current he could induce a greater loss of consciousness than produced by a continuous current alone. A 1 Hz current on its own had no effect on consciousness. Other research has found that adding a different type of field can also increase consciousness.[23]

Becker theorizes, "The concept of a dual nervous system with a primitive DC analog system and a superimposed, sophisticated, digital nerve-impulse system is strengthened by the observation of these ELF effects. The DC analog system is influenced by ELF fields. In fact, the interception of the natural ELF fields appears to be one of its function."

He says, "The capacity of ELF fields to alter consciousness and behavior indicates that the interface between the analog and the digital systems may be involved in some of the higher nervous functions that we have difficulty explaining with the hard-wired model. If the digital nervous system—by which we see, hear, smell, taste, feel, and more—is the child of a more primitive system by which we grow, heal, and obey the physical rhythms of our world, then there must be an intersection, a meeting place between the two.

"Is this the home of the mind, the site of memory, logic and creativity? We saw how we can consciously cross over this interface and gain access to our own DC systems in order to heal ourselves. By the same mechanism, we may gain control over our thought processes and behavior as well," he believes.[24]

Unfortunately, Becker, the late Burr, and these other scientists have not provided more details on how this electromagnetic field actually works in terms of the soul and spirituality.

None-the-less, this discovery of interacting dual electrical system provides a mechanism to understand how immaterial energy influences not only our minds, but also our whole bodies and lives—and maybe our spirituality.

8—Faking Spiritual Experiences

There is substantial evidence that we are on the right track here, as several scientists have used electromagnetism to "mechanically" induce spiritual experiences in people.

As long ago as the 1920s, scientists knew that shocking neurons in the brain could recall or generate a spiritual experience. Wilder Penfield found that by electrically stimulating the limbic cortex above the amygdala he could generate "spiritual" experiences in some subjects.[1]

More recently, Michael Persinger of Laurentian University in Ontario, Canada, found a way to reliably generate spiritual experiences in subjects in his laboratory by focusing electromagnetic fields at specific parts of their brains.

Persinger and his team use a helmet with solenoid coils to pulse electromagnetic fields into the temporal lobes of the subject. About 80% of volunteers who have worn the helmet have reported a presence of someone in the room with them, despite their being alone.

Persinger targets very weak currents, currents less than those in a hairdryer or a word-processor, to specific parts of the brain. The weak electromagnetic field imitates a complex, pulsing brain wave sequence known to produce LTP in the hippocampus. The pattern of the wave is more important than its intensity, he notes, citing the best results are if the magnetic fields are first focused over the right hemisphere for about 15 minutes followed by bilateral stimulation over the temporal-parietal lobes with a burst-firing field once every 3 or 4 seconds at 6.3–7.8 Hz, or even lower, and targeted predominantly at the amygdala and hippocampus.

What is experienced, he says, is "a synthetic ghost, a ghost generated by technology—in this case, the technology being the computer-generated electromagnetic fields through your brain. And the important thing is they were not very intense, They could be generated in nature, easily."[2]

Persinger hypothesizes these jolts of electromagnetism create micro electrical seizures or mini electrical storms that shake-up neural networks and reroute electric impulses in the brain that release a flood of images. The focused electromagnetic field somehow also entwines the subject's dream and waking consciousness, with the result that they are perceived in tandem, he suggests.

While the helmet generates spiritual experiences in most subjects, it does not create the same spiritual experience in every person. Everyone reports a different experience. "One crucial quality is the expectancy created by the setting," says Persinger. "In one study we played, in the background, the theme song from Close Encounters of the Third Kind; in another we had a cross hanging, slightly elevated, fifteen degrees to the left. Not surprisingly, the content reflected the setting. With the cross, in particular, there were these death themes, religious experiences….Then there was this transcendental meditation teacher. In the last two minutes of a twenty-minute session, she experienced God in the laboratory, visiting her. And afterwards we looked at her EEG, and there was this classic spike and slow-wave seizure over the temporal lobe at the time of the experience—the other parts of the brain were normal."[3] Overall, he found the experience of each subject reflects their beliefs and cultural background. The more religious the person the more religious the experience, while atheists tend to sense a presence or detachment from their bodies.[4]

These differences are believed to be due to different neural pathways, the different ways the subjects have experienced the world and tagged their learning, memories and emotions. This indicates that personal knowledge and experience are important in developing spiritual experiences untainted by cultural icons. (This very much reflects the approach to spirituality by Buddhism and some other Eastern religions.)

Persinger also notes that if the electromagnetic fields induce a sensation of a presence during stimulation of the right hemisphere, that presence is generally a more emotional or fearful experience. "If the experience does not occur until the bilateral stimulation, the location of the 'being' is perceived to be more to the right or on both sides and to be pleasant, even spiritual."[5] He points out how the right half of the brain is known to be responsible for creativity, while the left hemisphere is responsible for particular skills, such as language and logical thinking. When people experience uncertainty, the activity in the right hemisphere of their brain increases. In contrast, someone suffering pain or great stress may reduce activity in the left hemisphere, the location of self, allowing activity in the right hemisphere to dominate. This can result in increased dreaming, visions and even hallucinations. Persinger says, "The result of repeated right-hemispherical invasions of this kind is, paradoxically, to kick-start new activity in the left. And as the left hemisphere has this culture of optimism, the next thing the subject understands is this great surge of optimism, joy. All fearful, right-hemispherical effect falls away, confidence comes flooding in, and, if the conversionary process goes far enough, other left-hemispherical processes cut in too—an enhanced, rather than fragmented, sense of self—including self-esteem, a pressing need to structure all these brave new feelings into a system, otherwise a moral code, a determination to make sure everyone else feels just exactly the same way too."[6] This is one reason why stimulating one half of the brain more than the other seems to work better than stimulating both hemispheres.

When activity in the right hemisphere of the brain is out of synchronization with activity on the left, such as during a temporal lobe transient, Persinger says, the left hemisphere interprets the mismatched activity as another presence or what some say is God.

He theorizes that when you have a decrease in electrical activity in one important part of the brain with an increase in activity in another important part, what results is some form of religious experience. This is generally associated with intense stimulation of the subcortical areas of the temporal lobes, in particular the amygdala (which is associated with the sense of emotion) and the hippocampus (associated with personal memories). The associated excitation of these areas results in the experience being tagged with heightened personal meaning.

In short, Persinger suggests that this is one half of the brain being superaware of the other half, which it normally isn't.

Taking this a step further, is the soul then the whole brain being superaware of itself? Or are spiritual experiences reported by people over thousands of years due to just one part of the brain going quiet while other parts are overstimulated? Possibly. You be the judge.

Persinger and his team found the presence of a particular low-frequency component of a person's electroencephalogram, a theta rhythm, can reliably predict the likelihood of whether or not that person will have a religious experience.

He also notes how the "God experience" usually involves euphoric and positive emotions. "The person reports a type of God high that is characterized by a sense of profound meaningfulness, peacefulness and cosmic serenity. If the symbol is a father image, then the person expects to become a child of the father. If the symbol is 'imageless,' the person expects to become a part of the Universal Whole. Sometimes God Experiences can have negative emotional valences. During these periods, the same sense of oneness is pervaded by anxiety and fear.[7]

"These results strongly suggest that the experience of a "sentient presence" is an intrinsic property of the human brain and human experience." He sumizes, "The God experience is a normal and more organized pattern of temporal lobe activity."[8]

V.S. Ramachandran also believes electrical activity in the temporal lobes are key to spiritual experiences. He discovered that people with temporal lobe epilepsy produced much greater galvanic skin responses when looking at religious words compared to those who did not have this form of epilepsy. "What we suggested was that there are certain circuits within the temporal lobes which have been selectively activated in these patients and somehow the activity of these specific neural circuits makes them more prone to religious belief," he says.[9]

In similar research, Swiss neurologists believe the sensation of being outside your body is triggered by stimulation of another part of the brain, the angular gyrus in the right cortex.[10] Peter Brugger believes that when people "see" their body from outside it may be caused by over activity of the parietal lobes.[11]

Other studies suggest the sensation of leaving your body involves the hippocampus, which is intimately involved in spatial recognition and processing.[12] These sensations could be the result of a stochastic resonance in various parts of the nerve circuits, prompted by the electromagnetism.

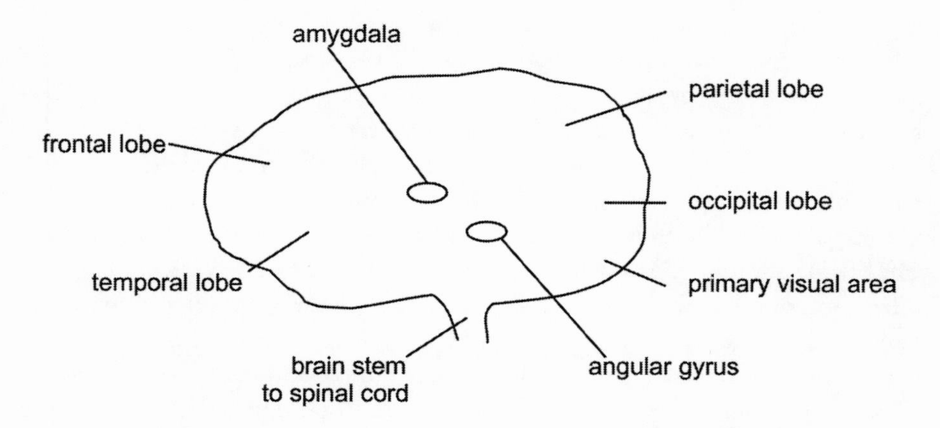

Figure 6. The human brain and location of key anatomical elements

Persinger notes "natural" spiritual experiences occur once every few years within the normal person. He suggests they are "precipitated by personal crises, such as the loss of a lovedone (real or imagined) or the confrontation of an insoluble problem, Certainly the greatest insoluble problem is the anticipation of self-extinction. Death anxiety increases in incremental steps as the person ages and approcaches the latter portion of life. God experiences proliferate during these periods and may even occur as death-bed episodes. The God experience is followed by a remarkable anxiety reduction and a positive anticipation of the future."[13]

These experiences can be triggered not only but electromagnetism, but also loss of oxygen to the brain (as in high altitude or near death experiences), changes in blood sugar (as in crisis situations, prolonged anxiety, fasting or other stresses) as well as some external electromagnetic effects.

For example, Persinger found that UFO sughtings were more frequent near fault lines in the earth's crust and that the number of sightings increase around the time of earthquakes. Fault lines can create substantial electromagnetic forces as rock crystals move against one another. He found these electromagnetic fields were similar to the ones he created to stimulate brains in the laboratory. As a result, he suggests that some people in areas around active fault lines and earthquake zones are likely to have strange and unfamiliar experiences, which they may interpret as encountering alien beings or UFOs.

Other research has also found increases in admissions of patients to psychiatric hospitals as well as increases in seizures of epileptics as a result of increased geomagnetic activity, such as earthquakes and tremors.

Persinger and his team also found an association between magnetic fields and reports of the supernatural, of ghosts and other paranormal experiences. They measured the magnetic fields in a small house where two adults lived, adults who reported occurences of apparitions, sensed spirits presence, nightmares and "waves of fear". The experiences often occurred within an area in which there were magnetic fields with irregular amplication between 1–5 microTesla (50mG) from 60 Hz sources. These amplifications were between a few seconds to just under a minute. This appeared to be generated by poor grounding of wiring in the house. Persinger suggests these magnetic fields evoke experiences in the brains of sensitive individuals. These people then use various cultural labels to explain and account for these experiences.[14] For instance, Persinger and his colleagues found a woman who reported nightly visitations by a spirit, a spirit she described as the Holy Spirit, appeared to be affected by a magnetic anomally generated by an electric clock about a foot from her head as she slept. The complex 4 microTesla magnetic pulses were similar to in frequency shape to those that evoke electrical seizures in epileptic rats and even humans.[15] They also tested this in the laboratory, where a subject reported "rushes of fear" and an apparition after being exposed to a weak, but complex 1mT magnetic field. ECG measurements were made at the same time and revealled temporary 1–2 second 15Hz spikes in his brainwaves that accompanied the sensation of fear.

In a supporting study, UFO researcher Albert Budden visited the homes of many people who reported they were abducted by aliens and found their houses to have unusually strong or abnormal electromagnetic fields. Abductees often report a feelings of paralysis, feelings of floating, bright lights and a sense of other beings nearby—feelings similar to those experienced by Persinger's subjects.[16]

This indicates that electromagnetic anomalies, either natural or the result of faulty electrical equipment, can be a source of "spirits" and spiritual experiences—a sort of "spiritual" playback device. This suggests "ghosts" are really just electromagnetic anomallies in the environment that trigger these images in our heads.

Persinger and a colleague even went as far to state there "is not a single case of haunt phenomena whose major characteristics cannot be acommodated by understanding the natural forces generated by the earth, the areas of the human brain stimulated by these energies, and the interpretation of these forces by normal psychological processes."

Interestingly, Persinger and his also team found the application of low-power magnetic fields of 10mT to the head can alter a person's perception of time. Other studies have shown that human cells can sense electromagnetic fields as small as a 10-millionth of a volt.

He suggests that besides creating changes in brain waves and electric circuits, focused electromagnetism also affects some of the brain's chemistry. "One of the likely neurochemical mechanisms by which increases in geomagnetic activity encourages electrical liability within the limbic system is the suppression of the nocturnal levels of melatonin. This serotonin-derivative, primarily synthesized within the pineal organ, has dampening effects upon hippocampal electrical activity. Decreases in melatonin levels have been correlated with periods when daily geomagnetic activity increases above 20 nT and have been evoked by nocturnal application of experimental magnetic fields with slightly larger strengths."[17]

The pineal gland, often described at the "third eye" by mystics, produces melatonin and serotonin, two neurohormones that control the body's biocycles, among other things. Research has found that by slightly increasing magnetism around the pineal gland can increase production of these neurohormones, while decreasing magnetism decreases production.[18]

Other researchers have found another magnetically sensitive chemical, magnetite crystals, elsewhere in the human brain. They are believed to be formed by biological processes, rather than contamination,[19] and enable the brain to detect small magnetic fields.

These chemicals are not found just in the brain. Much more magnetite (percentage wise) is found in animals that use the earth's magnetic poles for navigating, such as bees and certain species birds and fish. Pigeons whose eyes' are covered can still find their way home, however they cannot when small magnets are attached and interfere with the magnetite and their homing sense. Research has revealled a layer of cells containing tiny crystals of magnetite between the pigeons' brain and the skull, which allow the birds to find north accurately. Other research has found magnetic cells in bee's abdomen and a magnet placed in or around a their hive disorients them. Other animals with such sensivity to magnetism include bacteria, whales, turtles and butterflies.

As for those who claim special mystic powers to magnetism, there really is nothing mysterious about it. Magnetism is generated by the movement of an electric current and is therefore generally an accompanying effect of electricity—and creates an electromagnetic field. When an electron spins around the axis of an atom, the moving charge constitutes an electric current. A flowing current affects other electrons around it and creates magnetism. (A spinning electron acts like a minature bar magnet whose north and south poles are aligned along its axis of spin. Each electron is affected in only one of two ways—moving a specific distance up from its initial line of travel or a similar distance down.)

There are electric and magnetic currents all over the earth, between the earth's molten mantle and crust, between the crust and the ionesphere. These produce various currents and fields, such as telluric currents in the earth's surface, oloidal fields and geomagnetic impulses. (Overall, the earth's magnetic field fluctuates between 24–67 microtesla, pulsing at 7.8Hz and has a weak field of about half a gauss.)

The impact of magnetism appears to be limited on the human body, which appears to be quite used to the earth's magnetic field. Magnets do not bind iron in the blood. However, subjects who are insulated from magnetic fields do develop longer and more irregular biorhythms.

The 18th century French scientist Andre Marie Ampere, who discovered electromagnetism, described magnetism as "electricity thrown into curves". And this is just what magnets appear to do. Magnets can be used to divert the flow of energy.

A key point from this research is that directing any old electromagnetic wave into your brain is not necessarily going to generate a spiritual experience. For example, neurologist Mark George found that using transcranial magnetic stimulation focused on a similar area to that targeted by Persinger simply resulted in the jerking of the subject's thumb.[20] It takes very specific waves to generate spiritual experiences. In fact, some studies have failed to reproduce the results above because they have not accurately reproduced the correct form of electromagnetic waves and wave pattern. As mentioned earlier, waves have to be characterized in all their dimensions—by frequency, polarization and so on. Unfortunately many experiments just consider hertz, leaving other factors to chance—therefore, making them harder to repeat and to understand the results.

It is also interesting that many epileptics report spiritual sensations before or after a seizure. In fact, one woman reportedly did not want to be cured of her fits as she was afraid that she would loose her "connection" to God. Epilepsy is well known to be the result of abnormal electrical discharge in the brain. There is also some evidence that epileptics are more likely to have convulsions when there is an increase in the variation of the earth's magnetic field.

This would appear to reinforce Persinger's work that natural electromagnetic events, as well as those in his laboratory, can also prompt spiritual experiences. Was the inspiration of various prophets nothing more than a spark of mental electricity, or a quietening of their parietal lobes due to magnetically charged rocks? Is Persinger's helmut as the 21st century church. Is this the answer to our search? Are our spiritual experiences nothing more than electromagnetic waves resonating in certain parts of our brain? Or is this a "fake" experience, much like artificially colored of flavored foods?

Whatever the case, with this research we now know that spiritual experiences involve electromagnetism—a once mysterious force, but now well understood physical property of the universe in which live.

9—The Power of Belief

Interestingly, it is not just scientists that can create things spiritual, some people seem to be able to perform their own miracles.

In an intriguing experiment, one surgeon "operated" on 10 men to relieve arthritis pain in their knees. However, he only performed the standard arthroscopic surgery, scraping and rinsing the knee joint of two of the men. Three had the rinsing alone and five had no surgery at all, although the surgeon made small cuts in their skin to mimic incisions. None of the men knew about this and after the surgery all 10 men were provided with painkillers and crutches. Yet, six months later, all of them reported less knee pain![1]

Another study reported similar results, though in this experiment students who brushed harmless leaves on one arm were told the leaves were toxic. More than half of them developed a rash. However, other students who touched leaves that were toxic and were told they were safe had no reaction![2]

Similarly, psychologist Daniel Goleman reported a boy with a multiple personality disorder, where one of the personalities was allergic to orange juice, while the other personalities were not. How can allergic reactions be turned off and on at will? The white blood cells of the immune system are chemically triggered to react upon the contact of an antigen.

A further study showed that depressed patients who received a placebo showed brain wave activity in the same region of the cranium as patients who were administered a real anti-depressant.[4] Addressing the issue from the other side, Italian researchers found that when patients did not know they were receiving a medicine it did not work as well as when patients knew it was being administered. There are many other similar findings.

This placebo effect was defined in 1955 by Dr. Henry Beecher and remains a powerful if albeit little understood force, a mysterious empowering of mind over body. Scientific studies have shown that placebos can lower heart rates, relive pain, reduce kidney stones, alleviate asthma and lift depression.

Today, researchers believe that placebos work, for some people, through the ritual of being assessed by a physician, treated and receiving "medication" developing an expectation in the subject that they are being helped, that they are going to get better.

There appears to be similar mind-body action of people under hypnosis. For example, in a study of hypnotized subjects who were told a black and white photo was in color, their brain processed the image as if it was in color. Brain scans indicated blood flows in their brains appropriate to what the colors they thought they were seeing, not what they were really seeing. So, as far as their brains were concerned, the black and white pictures were in color, according to the researchers in California. In another case, the memory of people who believed they were drinking alcohol was impaired, even when they were just consuming water.[5]

This demonstrates how a body is not only influenced by thoughts, but can also change to meet those thoughts!

In short, the way you think sends information throughout your body, even overiding some of the signals that come back from your senses to your brain. Accordingly, 17th century philosopher Rene Descartes' reasoning "I think therefore I am" may not be too wrong. Descartes, in a lesser-known quote, also said, "my soul, by which I am what I am".

However, research indicates that only 10% of people are easily hypnotized and only half of us can be hypnotized at all. This could explain why placebos work for some people and not others: some people more easily respond to suggestions, while others do not and require more rigorous scientific proof to believe that something is occurring.

What is interesting for us in terms of our investigation into spirituality, is what triggers these brain signals to override the input from outside. Is it mere expectation or the work of a soul?

What is known, is that expectations and emotions appear to have a major role to play. Take the example of a loving kiss. Or is it just an exchange of breath and germs or much more than simple biochemistry?

10—Lessons of Love

Any discussion about the soul would not be complete without a consideration of love.

In the *Bible*, the Gospel of John defines "God is spirit" and "God is love".[1] This has prompted people to equate spirituality to love. Some people suggest emotions such as love are messages from our souls.[2]

Experiences of love do share many elements in common with spiritual experiences. For example, love can make us feel like we are part of something greater, prompt great creativity, even can change our whole life. With so many similarities, what can love tell us about spirituality?

While some people say "love" does not actually exist, that it is just chemicals, many people will testify to it from personal experience; just as others will as to spiritual experiences. Yet, it cannot be denied that "love," provides a major role in propagation and survival of the human species. Does spirituality do something likewise?

The physical aspects of love are revealing, in more than one sense. When you see or sense someone, chemicals don't easily cross the room and trigger a chemical reaction, there is something else initially at work. There appears to be something that initially causes the hormones, dopamine, norepinephrine and serotonin, to flow in us that ultimately lead to the powerful sensation of love. There is no doubting the power of thought, of brain waves, in love. Just consider how thoughts coupled with mechanical stimulation of genetalia and a flow of chemicals can lead to the thunder of physical orgasm.

Just looking at an image of a person can generate perceptions and electric waves that create feelings of love, lust or even hate.

This variety of responses is due to the references in the circuits of our minds. For example, a photo of a loved-one generally reignites harmonious and coherent waves, while an image of a naked person can either lead to lust or repulsion depending on the reference of nudity: a physical education teacher might have said it is natural, while a preacher may have informed you that nudity is akin to the devil. Which perception is stronger is reliant upon which lesson impacted you the most at the time or how you have developed your knowledge and frames of reference since then. And as we know, perception involves electric currents and circuits. Interestingly, this is more than just a visual recall, as a drawing or cartoon of a loved one or naked person can also generate the same response.

When it comes to love, focusing perceptions focuses electric waves, creating coherently resonating waves in our heads and bodies. Physical stimulation and/or waves from a loved one can add more waves, maybe in the form of a stochastic or other resonance, that ratchets the power of these waves even higher to provide the experience of love. Chemicals keep these waves flowing, even intensifying them. At some stage, the waves trigger or break through a threshold and the feeling of love or thunder of organism is unleashed,with an asssociated flow of chemicals amplifying the affect.

In contrast, jagged waves can be made further chaotic by other perceptions and emotions. In this instance, the process results in a downward spiral, with waves being amplified more chaotically and stress being placed on the nervous system and body, and fear amplified, if no release is possible.

The impact of emotions on our nervous system is being revealed in groundbreaking research by the Institute of HeartMath in California. It notes how the actual number of neural connections going from emotional centers to the cognitive centers is greater than the number going the other way. "This goes someway to explain the tremendous power of emotions, in contrast to thought alone."[3]

The institute notes, "Thoughts and emotional states can be considered 'coherent' or 'incoherent.' We describe positive emotions such as love or appreciation as coherent states, whereas negative feelings such as anger, anxiety or frustration are examples of incoherent states....different emotions lead to measurably different degrees of coherence in the oscillatory rhythms generated by the body's systems. This leads us to a second use of the term 'coherence.' In physics, the term is used to describe two or more waves that are phase—or frequency-locked together to produce a constructive waveform. A common example is the laser, in which multiple light waves phase-lock to produce a powerful, *coherent* energy wave. In physiology, the term is similarly used to describe a state in which two or more of the body's oscillatory systems, such as respiration and heart rhythm patterns, become synchronous and operate at the same frequency. This type of coherence is called *entrainment.* The term coherence is also used in mathematics to describe the ordered or constructive distribution of the power content within a *single* waveform. In this case, the more stable the frequency and shape of the waveform, the higher the coherence." The institute also notes: "We have found that in states in which there is a high degree of coherence *within* the HRV [heart rate variation] waveform, there also tends to be increased coherence *between* the rhythmic patterns produced by different physiological oscillatory systems (e.g. synchronization and entrainment between heart rhythms, respiratory rhythms and blood pressure oscillations)."

It adds, "the key to the successful integration of the mind and emotions lies in increasing the coherence (ordered, harmonious function) in both systems and bringing them into phase with one another. It is our experience that the degree of coherence between the mind and emotions can vary considerably. When they are out-of-phase, overall awareness is reduced. Conversely, when they are in-phase, awareness is expanded. This interaction affects us on a number of levels: vision, listening abilities, reaction times, mental clarity, feeling states and sensitivities are all influenced by the degree of mental and emotional coherence experienced at any given moment," the institute reports on its website.[4]

Meditating about love, compassion and caring generate greater coherence in your heart waves (ECG) than people who are simply resting. Rollin McCraty of the institute found that during feelings of love, heart rhythms displayed ordered, smooth and harmonious sine wave patterns. "Many different organs and systems contribute to the patterns that ultimately determine our emotional experience. However, research has illuminated that the heart plays a particularly important role. The heart is the most powerful generator of rhythmic information patterns in the human body. Our data indicate that when heart rhythms patterns are coherent, the neural information sent to the brain facilitates cortical function. This effect is often experienced as heightened mental clarity, improved decision-making and increased creativity. Additionally, coherent input from the heart tends to facilitate the experience of positive feeling states. This may explain why most people associate love and other positive feelings with the heart.....the heart is intimately involved in the generation of psychophysical coherence," they say.

McCraty also says these heart waves are a key to our emotions, thinking and actions. "Numerous experiments have now demonstrated that the messages the heart sends the brain affect our perceptions, mental processes, feeling states and performance in profound ways. Our research suggests that the heart communicates information relative to emotional state (as reflected by patterns in heart rate variability) to the cardiac center of the brain stem (medulla), which in turn feeds into the intralaminar nuclei of the thalamus and the amygdala. These areas are directly connected to the base of the frontal lobes, which are critical for decision-making and the integration of reason and feeling. The intralaminar nuclei send signals to the rest of the cortex to help synchronize cortical activity, thus providing a pathway and mechanism to explain how the heart's rhythms can alter brain-wave patterns and thereby modify brain function. Our data indicate that when heart rhythm patterns are coherent, the neural information sent to the brain facilitates cortical function. This effect is often experienced as heightened mental clarity, improved decision-making and increased creativity. Additionally, coherent input from the heart tends to facilitate the experience of positive feeling states. This may explain why most people associate love and other positive feelings with the heart and why many people actually "feel" or "sense" these emotions in the area of the heart."[5]

It is not just the heart that is more electrically active during emotional states, so is the brain. LeDoux notes more brain systems are typically active during emotional than nonemotional states, and when the intensity of arousal is greater the opportunity for coordinated learning across brain systems is greater during such emotional states. For example, you jump back from a car as it swerves toward you and your senses instruct your muscles to react, then you consciously realize you have just jumped back as the information reaches your brain and perceives and processes it and then you feel afraid as the chemicals rush through your body and emotions do their thing to attach to the memory that is now forming.

Emotions help amplify memories, somehow attaching to a LTP as it is learning and forming a memory.[6] This process is known to involve the amygdala, which can tag an ordinary experience or perception as a highly emotional one.

The Institute of HeartMath says, "The amygdala compares incoming emotional signals with stored emotional memories. In this way, the amygdala makes instantaneous decisions about the threat level of incoming sensory information, and due to its extensive connections to the hypothalamus and other autonomic nervous system centers, is able to 'hijack' the neural pathways activating the autonomic nervous system and emotional response before the higher brain centers receive the sensory information. One of the functions of the amygdala is to organize what patterns become 'familiar' to the brain. If the rhythm patterns generated by the heart are disordered and incoherent, especially in early life, the amygdala learns to expect disharmony as the familiar baseline; and thus we feel 'at home' with incoherence, which can affect learning, creativity and emotional balance. In other words we feel 'comfortable' only with internal incoherence, which in this case is really discomfort. On the basis of what has become familiar to the amygdala, the frontal cortex mediates decisions as to what constitutes appropriate behavior in any given situation. Thus, subconscious emotional memories and associated physiological patterns underlie and affect our perceptions, emotional reactions, thought processes and behavior.

"In sum, from our current understanding of the elaborate feedback networks between the brain, the heart and the mental and emotional systems, it becomes clear that the age-old struggle between intellect and emotion will not be resolved by the mind gaining dominance over the emotions, but rather by increasing the harmonious balance between the two systems—a synthesis that provides greater access to our full range of intelligence," the institute proposes.[7] It has developed several effective techniques and tools to help people do this.

Emotionally-influenced memories are processed differently in the amygdala of men and women, according to research by Richard Davidson. He found that in women, emotional memories are processed on the left side, while in men they are processed exclusively on the right side.[8]

In periods of intense stress, memory functions of both women and men are weakened with it is harder to recall memories as well as develop new ones. Research has found that stress impedes long-term potentiation in the hippocampus and can even shrink, with the size of the hippcampus smaller in chronically depressed patients compared to those who are not, or who have recovered from depression.

In contrast, memory is enhanced by coherent waves. But why does one brain organ respond negatively to this and another positively? Further research is obviously required.

One reason might be that the amygdala and hippocampus mediate different kinds of memory, as notes LeDoux. The way these two organs work together is so close that emotional memories (mediated by the amygdala) and memories of emotion (mediated by the hippocampus) can not be easily separated.[9] He also suggests the amygdala can influence working memory through altering sensory processing in cortical areas or acting directly on working memory circuits. LeDoux even believes the amygdala can alter consciousness, transform cognition into emotion, and even allow emotion to overtake consciousness. Emotion can monopolize consciousness, at least when it comes to fear, with the amygdala dominate working memory, he says.[10]

Other studies on the effects of stress on the brain indicate that stress and conflict impair memory by altering the functioning of the hippocampus.[11] In addition, stress also seems to enhance the amygdala's response to fear.[12]

This acting on two fronts, enhancing the amygdala's response and altering the hippocampus, can make a perception of fear seem worse than it really is. This suggests the greatest fear we can have is simply of fear itself.

A spiritual experience appears to some degree to invoke the opposite. For example, researchers suggest that in some instances, a perception is heightened by the hippocampus, while the amygdala attaches a greater than normal emotional response to it. Again, this acting on two fronts could make a perception more spiritual than it really is.

One of the roles of the hippocampus is to convert short-term memory to long-term recollections, according to neurobiologists at the University of Amsterdam. The hippocampus contains two sorts of brain cells: pyramidal cells and interneurons. The pyramidal cells process incoming information and pass it along to other pyramidal cells in other parts of the brain, such as those in the cortex involved in LTP. In contrast, the interneurons can inhibit the pyramidal cells and control the information they are able to receive and pass on. Just how they make these decisions is uncertain.[13]

Other researchers believe the whole limbic system, which includes the amygdala and hippocampus, is a convergence zone of nerves, becomes unusually active during a spiritual experience attributing everything as having special significance.

Limbic system structures such as the amygdala, hippocampus, and the inferior temporal lobe have been shown to provide the foundations for mystical, spiritual, and religious experience, and the perceptions, including the 'hallucination' of ghosts, demons, spirits and sprites, and belief in demonic or angelic possession, notes Rhawn Joseph in his book *Transmitter to God*. "The amygdala enables us to hear 'sweet sounds,' recall 'bitter memories,' or determine if something is spiritualy significant, sexualy enticing, or good to eat. The amygdala also makes it possible for us to store personal, sexual, spiritual, and emotional experiences in memory and to recall and reexperience these memories when awake or during the course of a dream in the form of visual, auditory, or religious or spiritual imagery. The amygdala, in conjunction with the hippocampus, contribues in large part to the production of very sexual as well as bizarre, unusual and fearful mental phenomen including out-of-body dissociate states, feelings of depersonalization, and hallucinogenic and dream-like recollections. The amygdala also makes it possible to experience not just spiritual and religious awe, but all the terror and dread of the the unknown. Indeed, the amygdala can generate feelings of hellish, nightmarish fear. And yet, it is also the amygdala which is responsible for the capacity to transcend the known, this reality. It is also the amygdala which assists in maintaining this reality through the inhibition and filtering of most of the sensory signals bombarding the brain and body....Neurosurgical patients have reported communing with spirits or receiving profound knowledge from the Hereafter, following depth electrode amygdala stimulation or activation."

Joseph adds that whereas the amygdala and hypothalamus interact in regard to pleasure, rage and sexuality, the amygdala and hippocampus interact to subserve and mediate different aspects of experience, including memory, dreaming and hallucinations. He says the hippocampus, in particular, appears to be responsible for certain types of 'halucinations' such as the visualizations of astral projection or seeing oneself floating above the body. Joseph also notes that in response to pain, fear and terror, the amygdala and other elements of the limbic system begin to secrete opiates which eventually induce clam, as well as analgesia and euphoria. "It is this heroin high that explains why an ox, deer, and other animals will simply give up and lie still while they are devoured and eaten alive," Joseph suggests.[14]

In contrast, LeDoux says that while nerve circuits incorporating emotions of fear and love both involve the amygdala, they are quite distinct circuits within it. Different stimuli and perceptions are also processed in other, different parts of the brain. LeDoux believes different systems of the brain are involved in different kinds of emotions.[15] This theory is supported by other research, such as that at the University College in London, which suggests the medial insula, the anterior cingulate cortex and striatum are also involved in the electronic circuitry of love, becoming active when people look at pictures of their loved ones.

Is there then another circuit for spirituality?

Most likely. The amygdala, or another part of the limbic system, could easily contain another circuit of nerves that tags various learning and memories with another "emotion," an emotion that we call spirituality.

While most of the nerve circuits in the amygdala operate independently in normal individuals, there are some indications that they can become crossed, or resonate at the same time.[16]

Love, lust and fear can share similar associated emotions, including spirituality. For instance, great love provides a feeling of great power, connectivity and a sense of purposefulness. In terms of great fear, I recall rock climbing and reaching my mental and physical limit (regarding what I thought I could do) and was about to fall, at that point having a spiritual experience, an experience of connectedness with the universe and creator force. Such experiences suggest our spiritual circuits are parallel or close to those of love, lust and fear in the amygdala.

This closeness and interaction of emotions could explain how some perceptions and emotions appear to become crossed. As many of us have noticed in our own lives, emotions not only help us make decisions they can also interfere with logic.

It could also explain how great love is and spirituality is sometimes accompanied by great hatred—and a range of conflicting thoughts and actions, such as killing in the name of God.

Interestingly, it appears that the female limbic system is slightly different from that of males. For example, there are more neurons in a woman's amygdala and they are more closely packed, notes Joseph. He adds the anterior commissure of nerve fibers which connects the amygdala with the right and left temporal lobes is also 18% larger in women than in men.[17]

To better determine if spirituality does involve the limbic system, we have to better understand the limbic structures.

The limbic system, sometimes called the "emotional brain," is a complex and weird looking collection of structures deep inside our heads. This collection of densely packed nerves, which include the amygdala, hippocampus, hypothalamus and thalamus, plays a primary role in interpreting pain, pleasure and the range of human emotions and associated physical responses. Besides this, it is also known that the limbic system plays a role in storing short-term memory, as people with damaged limbic systems forget recent events and cannot commit anything to memory.

An interesting, but often overlooked, part of the brain adjoining the limbic system is the cerebrospinal fluid ventricles, which are fluid-filled cavities around the limbic system. They are believed to act as a form of cushioning, but may also assist in amplifying or dampening brain wave resonance. These fluid-filled cavities also maintain the optimal ion and chemical environment for neuronal signaling. Just a slight change in the ionic composition of this fluid can seriously disrupt nerve signaling. This fluid also circulates throughout the brain and spinal cord, carrying ions, chemicals, glucose and oxygen from the blood to neurons, acting like a "blood" of the nervous system.

It is also important to note that besides the electromagnetic signaling of nerves in the limbic system and elsewhere that there is an ensuing range of electrochemical and then biochemical actions and reactions. This is particularly strong when it comes to emotions. Remember the work of Pert and her molecules of emotions. These, and other chemicals, are believed to "tag" emotions by attaching to, and altering the synapses between neurons. This means that they then only respond to a certain emotion, or that when they do respond or remember, that the emotion is also remembered. For example, the chemical compound oxytocin, which is often referred to as the "love hormone," is known to be important in childbirth and generating breast milk, and has now been found to be involved in learning—and therefore has some association with LTP. The Institute of HeartMath has found there is more oxytocin in the heart than there is in the brain. Further research will undoubtedly shed more light on this fascinating area and the range of opportunities for human health—as well as for pharmaceutical companies.

Yet, while science notes that "love" is accompanied with fascinating chemical reactions, little study has been undertaken into the electromagnetic elements underlying those chemical interactions, and which may well be primary to them.

11—The Evidence So Far

When all of these discoveries and developments are viewed together we get a good idea of how the soul might operate and spiritual experiences generated.

Basically, every second of every minute of everyday we are alive we balance myriad internal and external demands. This is accomplished, often unconsciously, by the nervous system, predominantly the autonomic nervous system. The autonomic system comprises two circuits, the sympathetic and parasympathetic nervous systems. These systems usually present two opposing courses of action to each stimulus presented to us. For example, the parasympathetic nervous system can increase the heart rate while sympathetic nerves can decrease it—the former prompting us to get excited and ready to act, while the latter calms us. The autonomic, and rest of our nervous, system resolves and regulates the information coming in from these two nervous systems, references it, make sense of it and acts upon it.

All of this information is conveyed by two electric systems, digitally pulsing nerve cells and continuous current perineural cells. These two systems provide greater information and processing capability than just one system—and provides a way to create holistic perceptions from energy (as we will see later).

If the information sensed is new and/or important enough, these electric currents potentiate and resonate around nerve circuits.

If strong enough, this potentiation entrains other electric waves in the brain and ultimately alter synapses between nerves and create new circuits that help us perceive, learn, memorize and recall.

This long-term potentiation, or LTP, of resonating nerve circuits and brain waves occurs predominantly in our heads when confined to the operation of the mind.

When it comes to spirituality, it is more than likely that this resonance also occurs in various parts of the body, enjoining and entraining other electric currents such as heart waves and more. This could provide a basis for, and explain, the greater power of spiritual experiences. Spiritual experiences are LTPs that involve a coherent resonance of brain, heart and other waves rather than just predominantly brain waves.

This more extensive LTP is also likely to involve stochastic resonance, whereby these coherent waves become even more sensitive to others around them, detecting things previously undetectable or unconscious to us.

This process is also ensued and accompanied by an intricate biochemical dance.

While it is known this process involves the brain's limbic system, which appears to decide what we need to be consciously aware of, it also involves emotional elements, tagging what we perceive: sometimes perceptions are labeled with fear, love or as something spiritual.

However, this does not necessarily mean that the limbic system in our brains is the long sought-after location of the human soul. Remember, coherence of brain and heart waves, as well as emotional tagging occurs in other parts of the body.

Rather, the soul appears to more holistically involve an integrated system of energy flows. There are likely to be other elements that we have not yet identified.

More abstractly, the soul is what puts together the lessons of our lives, suggests former Monk turned author, Thomas Moore. He describes the soul as "not a thing, but a quality or a dimension of experiencing life and ourselves. It has to do with depth, value, relatedness, heart and personal substance."[1]

What differentiates a spiritual experience from other conscious experiences involves:

- parts of the electromagnetic nerve and perineural circuits of the brain and body resonating together, and
- nerve circuits in the amygdala being influenced by this resonating electromagnetism, and/or
- the amygdala possibly attaching a greater than normal emotional tag to a perception heightened by the hippocampus, or
- a freeing of consciousness from these emotional tags, and/or
- one part of the brain and thought dominating others.

While the evidence is not conclusive, Persinger's experiments have shown that a targetted pattern of low-power electromagnetic waves and fields are a major component of spiritual experiences.

The experiments regarding suggestion, hypnosis and the effectiveness of placebos indicates that some people are able to influence their own brain and body by accepting, extending, or believing what they are told—creating their own "minor miracles" and shows that this again has something to do with brain wave patterns.

Together, these findings demonstrate there is a primary element of energy in the human mind, to human consciousness—what many people would call the soul. Each moment, disparate functions of your brain, nerves and energy are joined together to accomplish a transcendent task that no single part could do on its own. Immaterial electromagnetic energy is combined with impartial atoms, nerve cells and chemicals to create such subjective things as human consciousness, emotions, character and spirituality. Our souls and spirituality are much larger than the sum of their components.

Many scientists, and even theologians, acknowledge the soul and energy are related. Psychologist Carl Jung, for example, said "in physics, we speak of energy and its various manifestations, such as electricity, light, heat, etc. The situation in psychology is precisely the same. Here, too, we are dealing primarily with energy, that is to say, with measures of intensity, with greater or lesser quantities. It can appear in various guises. If we conceive of libido as energy, we can take a comprehensive and unified view. Qualitative questions as to the nature of the libido—whether it be sexuality, power, hunger or something else—recede into the background....I see man's drives, for example, as various manifestations of energetic processes and thus as forces analogous to heat, light, etc." Interestingly, in another instance he states, "I have been convinced that at least a part of our psychic existence is characterized by a relativity of space and time. This relativity seems to increase, in proportion to the distance from consciousness, to an absolute condition of timelessness and spacelessness."[2]

It is important to note that this energy, our nerves and brains are simply the hardware through which spirit is experienced. To say the physical brain alone produces spirituality is like saying a piano produces symphonies. In another analogy, you generally cannot sense your little toe and mostly forget why you even have them. But you can sense them if you focus on them—and it is then that you realize they are there to help you balance and walk. It is similar with our souls and spirituality; we tend to ignore them. But when we focus on them we can sense them and that they provide a function in our lives, even if we are not exactly sure just what that is yet.

Just how can the combination of electromagnetism and biological wiring of 100 billion neurons in the three pounds of your brain create a soul and spirituality? To better under this we have to better understand energy.

12—Equating Energy

Modern physics is revealing some amazing information about energy waves, and fields; information that can explain how resonating brain waves give rise to not only consciousness, but also to spirituality.

To understand energy and what it means to our soul we need to consider Albert Einstein's famous equation, $E = mc^2$ (where E = energy, m = mass or weight multiplied by c squared = the speed of light, 186,000 miles a second). This equation shows how energy has mass, and how mass has energy; even that matter is energy and vice versa. For instance, as electrons speed-up and gain energy, the Energy side of the equation increases. And since the speed of light is constant, the c^2 part of the equation cannot change, which means that the mass side has to change. The mass of an object increases, slowly at first, but increasingly quickly as it energetically accelerates nearer and nearer to the speed of light.[1]

Not only is mass equivalent to energy, if enough energy is concentrated the result is matter, such as the chair you are sitting on.[2] This occurs as particles of energy combine together to form sub-atomic particles, which in turn comprise atoms, which in turn make up material. Accordingly, energy is a fundamental component of all matter.

Is this physics how reverberating energy in our nerves creates new synapses in our brains? The answer is uncertain and requires further interdisciplinary research.

It is known that physics can be used to describe how LTP may be initiated, continued and even strengthened in our nerve cells. Remember how charged ions are the batteries of our cells: if an electric field is directed at an ion, that particle will begin to rotate at an angle to the applied field. The speed at which the particle spins is determined by the ratio between the mass of the particle and the strength of the field. If another electromagnetic field is added that has the same frequency as the frequency of the particle's rotation, energy is also added from the second field to the particle. (The second electromagnetic component could be generated by the perineural cells that surround the electric current of nerve cells.)

This adds to the strength of the particle/ion/electron's orbit, increasing its energy and affects on nearby particles in nerve cells.

And as electrons are sped-up they gain energy, which can in turn be used to strengthen the reverberations and the LTP process.

This field enhancement process has also been discovered to allow ions to pass through cell membranes more effectively and in larger numbers. This in turn increases the strength of the electric signaling from the cells. And as more particles are enjoined it creates cellular resonance and ultimately LTP.

These principles of energy acceleration and enhancement go some of the way towards explaining how energy in our heads can ultimately alter nerve synapses and circuits that are involved in creating learning and memories—in short, how we get thoughts and actions out of brain waves in our heads and bodies.

In contrast, when matter is destroyed it is released as energy, such as fire, radioactive decay and so on. In terms of our investigation, this could mean that an energy-based soul has additonal mass compared to a life-less brain. If energy has departed the body as in a just deceased person, it should be possible to detect a difference in weight, according to $E = mc^2$. Yet there appears to be no scientific research on the weight of a body just before and after death. However, there are reports that Polish physicist Janusz Slawinski recorded flashes of light at the moment of death of people. Slawinski describes the 'death flash' as 10 to 1,000 times stronger than the light emanating from the body when it is alive.[3] Also, people who have come back to life after being declared clinically dead report common perceptions of bright lights. Hunt suggests death is a transition from one vibrational energy state to a higher one.[4]

The physicist James Clerk Maxwell, whose equations provided a greater understanding about electromagnetism, believed that light was a fusion of traveling electric and magnetic fields. He also discovered that electromagnetism travels as waves at the speed of light (when traveling through space).[6] Waves of light are comprised of photons, the very same photons that are believed to convey signals between electrons.

When energy changes, as when a solid turns to a liquid or vice versa, there is what is called a "phase shift" in the atomic field. For example, H_2O can appear to us in a range of identities—be it ice, water or steam, but it is always H_2O.

At this point, it should be noted that $E = mc^2$ is not the complete Einstein equation, there is also a constant called the Lorentz constant or transformation, which sits underneath the famous formula. The Lorentz constant indicates that energy and matter are one related by motion transformation. This implies that a certain motion is required for a change or phase shift between energy and matter. Further research is required in this respect.

Interestingly, Frank Schubert says Einstein's formula suggests "the traditional barrier between God and man is replaced by an essential 'relatedness'. It is clear the Einstein's theory provides this very relatedness…for it seems to point to an equation which can, at the same time, account for energy and matter as true functions of each other ($E = mc^2$). A God of pure energy, then could well become 'matter' in an incarnation."[5]

This raises another interesting property of energy, the fact that it cannot be destroyed. This sounds analogous to a property of the immortal soul described in many religious texts.

13—The Wonders of Waves

While we have seen that electromagnetic waves can be used to generate spiritual experiences in people, just how can apparently simple waves result in something so complex?

A wave is a distortion in a medium where the individual parts of the material only cycle up and down or back and forth while the wave itself moves through the medium. It appears that something is moving along, but it really just is the distortion, or more accurately information, moving along the medium and influencing the next part. Also, today's quantum physics represents particles as both waves and particles (in contrast to classical physics, which described their actions in terms of tiny billiard balls bouncing around.).

There are different types of waves:

- transverse waves—are created by up and down motion. These types of waves include a "wave" of standing and sitting people at a ballgame or along a piece of rope or a guitar string. The particle motion is perpendicular to the motion of the wave;

- compression or longitudinal waves—created by back and forth motion, such as sound waves or in alternate current electricity (where electrons move back and forth, compared to direct current where they flow through the conductor). The particle motion is parallel to the wave motion; and

- circular waves—where material moves in a combination of transverse and compression (often a circular or elliptical pattern). This includes surface waves on top of water, where molecules travel in a circular motion, the crest being the top of the circle. Waves under the surface are longitudinal in nature.

Waves are characterized by wavelength, amplitude, velocity and frequency. Wavelength is the distance from one wave crest to the next. Amplitude is the height of the wave and relates to loudness in sound and brightness in light. Velocity is how fast a crest is moving from a fixed point. Frequency is the rate of crests passing a given point (or wavelength divided by velocity) and is given as peaks, or cycles, a second (also called Hertz).

The waveform and its information can extend over large distances, even through matter. (In the case of very low temperatures, what is called Bose-Einstein condensate occurs where the waves of atoms overlap and atoms lose their individual identities and essentially flow as a single wave with uniform behaviour.)

Virtually all forms of energy are propogated by waves.

Electromagnetic waves, which are produced by the vibration of electrons, are believed to be transverse waves, with up and down motion of electric and magnetic fields, perpendicular to each other. However, some scientists think electromagnetism and light may also be circular waves.[1]

Interestingly, electromagnetic waves do not require a medium for the wave to travel through!

Ultimately, energy does not require a material or medium to travel through or be stored in. Other waves, such as sound waves, can not travel through a vacuum, while electromagnetic and light waves can.

While electromagnetic waves carry no mass, they can carries energy and can exert pressure known as radiation.[2] This is how light can affect things, such as plants and ourselves.

We don't notice the wave aspect of energy and matter in our daily lives as the smallness of electrons obscure the wave-like aspects of matter; but they are there none-the-less. For example, visible light ranges from 400 nanometres and a frequency of 7.5×10^{14} Hz for reddish light to 700 nanometres and a frequency of 4.3×10^{14} Hz for violet blue light. Below 400 nm is ultraviolet light, followed by even shorter x-rays and gamma rays. Just above the visible spectrum is infrared light, then even longer radio waves, some of which can be measured in distances of meters. Why don't all these interacting electromagnetic waves just make a mess of our thoughts and bodies? Besides the earth's atmosphere absorbing many of those resonating from space, we appear to enjoy a form of insulation or relativity that stops low frequency electromagnetic waves from distorting those in our heads. This protection maybe provided by the perineural cells and current that encases our nerves—this is unless you wear one of Michael Persinger's helmets!

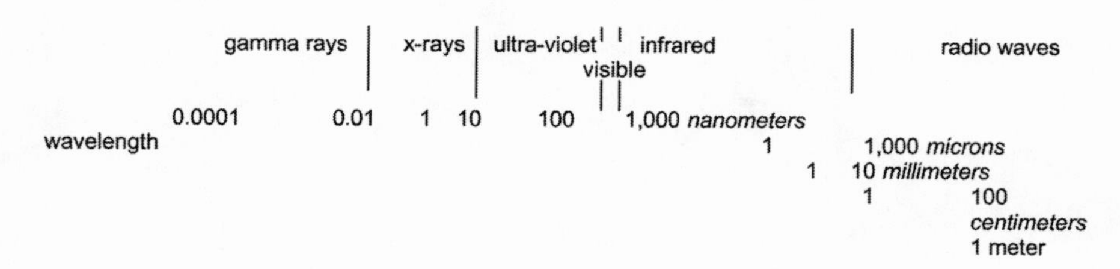

Figure 7. The wave spectrum

When waves are described mathematically a complex set of numbers is required to account for their amplitude, frequency, velocity, wavelength, shape and so on. This is in contrast to the on-off/0–1 electric impulses of nerve cells and computer circuits. These complex sets of numbers and shape of waves allow for much more information to be conveyed in each wave. Then there is the variation between waves which can also convey even more information.

While there are myriad wave shapes, it is the way that waves interact that is most important to our investigation.

According to scientific principles, waves interact with each other through the overlapping of their crests and troughs of these waves. This point is worth repeating, as it is the interaction and overlapping of waves that can create new waves (called constructive interference, coherence or entrainment) or cancel them out (destructive interference). For example, the wave from a rock dropped into a pond sends out orderly circular ripples. When you throw a second rock into the pool some of the crests of the waves add together to create larger waves, while others' troughs cancel others' crests out, leaving flat water.

When you throw lots of pebbles into a pond you get much more complicated action, with the mix of waves adding and subtracting in a complex manner. Sometimes there are so many waves that they interact like waves in the open ocean, creating choppy seas and breaking into white frothy tops here and there. (Interestingly, these frothy wave tops give the appearance of particles when viewed from a distance). And this involves basically only two dimensions; it becomes more complex when you involve three dimensions and time.

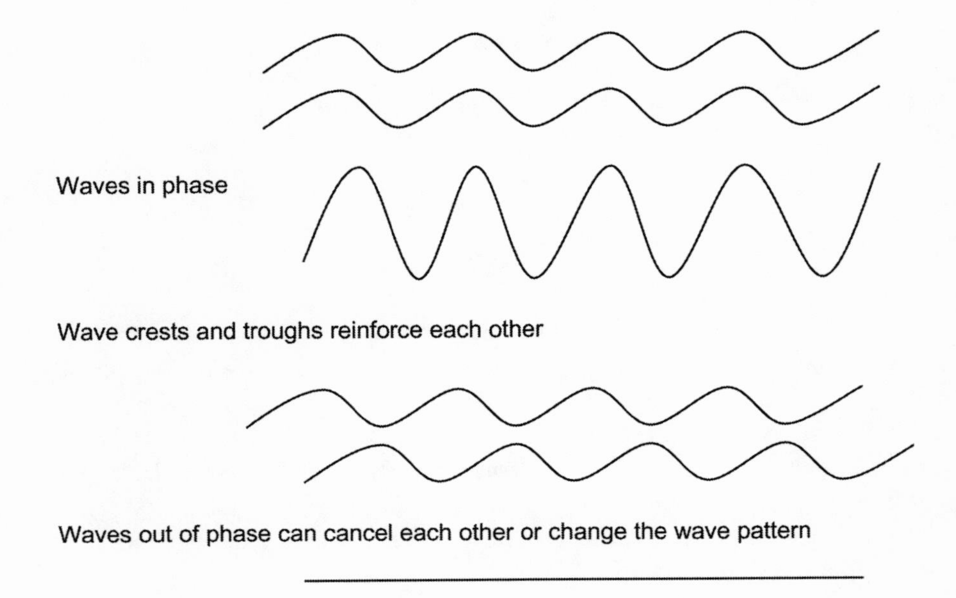

Figure 8. Wave interactions

In another example using light, remember those school experiments where a beam of light was projected through two narrow slits in a piece of paper onto a background and where rather than two bands of light, an interference pattern of alternating light and dark bands, like zebra stripes, appear.[3] These bands are the results of waves of light interfering with each other, much like ripples of water in a pond from two pebbles dropped into it. While each particle of light goes through just one slit, its motion is affected by the information contained in the waves going through both slits. Using a light with a higher frequency results in more bands due to the interference of shorter wavelengths.

Other researchers have found a similar affect can be achieved by bouncing electrons off nickel crystals. There is also apparently another crystal that absorbs one photon and emits two, albeit lower energy, photons.[4]

The theory of quantum electrodynamics notes that the way electrons circle around the nucleus of an atom is more wave-like, rather than orbital. So rather than being like the moon orbiting around the earth, the polarized electrons travel along a wave-like path weaving in and out. As they do so, electrons also send out and receive messenger photons, which appear to 'sense' what is around them, in particular the proton of their atom's nucleus. These messenger photons exhibit even more pronounced wave characteristics.

The composition and interaction of these electrons and photons, and their various wave motions, account for all the known atoms, chemicals and substances in our world. "All atoms, are made up of a certain number of protons exchanging photons with the same number of electrons," said Richard Feynman in *QED: the Strange Theory of Light and Matter*. "The patterns in which they gather are complicated and offer an enormous variety of properties." Accordingly, the interacting waves of these basic particles is fundamental to the world around us as well as out life.

Interestingly, photons and light energy takes the shortest path in terms of time, not the shortest path in terms of distance. This means that light does not always travel in a straight line. Time is the primary dimension influencing light. However, the speed of light can be slowed as it passes through glass, water and other materials as it interacts with their atoms. In fact, it has been almost stopped by super-cold sodium ions.[5] When light passes through these ions it interacts with the spin of the electrons of the sodium atoms, which slows its progress. It is believed that during this interaction, the photons of light disappear and the information they contained is stored in spin states of the sodium ions.[6]

Electromagnetic waves (outside the body) are particularly fascinating as while they can cancel each other out, which gives the appearance that the current is turned off, information continues to flow. This means the information contained in the original waves is not destroyed by the interference, but continues in terms of potentials. These potentials are known as the Aharonov-Bohm effect. Not only have this been proven, it is also used in many everday electronic conveniences.

Physicists point out that waves are also a form of information. For example, the way an electron vibrates provides information to those around it. In fact, every atom vibrates at specific wave frequencies. These frequencies can be detected from distances of billions of light years thanks to radio and other telescopes. This is one reason how scientists know what stars in distant universes are comprised. And no matter how buried a single atom may be in a mass of matter, its waveform continues to exert influence, albeit it often masked by the waveforms of the matter around it. Bohm provides greater detail (including numerous mathematical equations) about this process and wave interaction in his study *The Undivided Universe*.[7]

These vibrating waves are not just restricted to atoms. Physicists believe that vibrating "strings" of energy are the essence of all matter—of everything in the world. So not only do we have waves in our brain, hearts and bodies, but we are ultimately comprised of minute vibrating waves of energy.[8] If this string theory is correct, matter and energy, mind and spirit are not really all that different—they just appear to be to us.

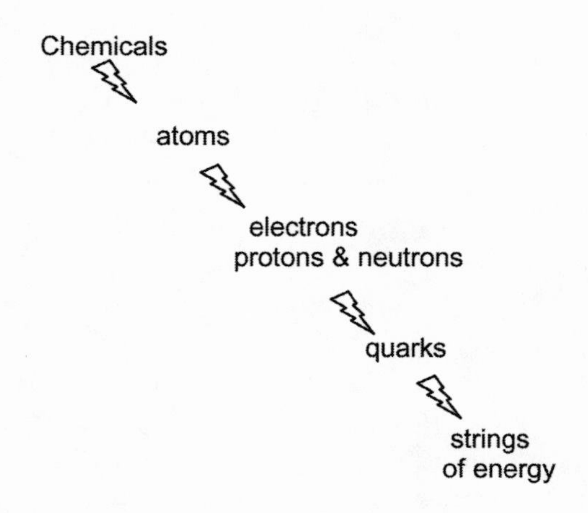

Figure 9. Beyond atoms are stings, vibrating strings of energy

In string theory, every particle of matter and every transmitter of force consists of a string, whose pattern of vibration is its fingerprint, notes Brian Greene in his book *The Elegant Universe.*[9] Strings are not particles that occupy a point of space, but rather have a single energy dimension (but no other dimensions) that occupies a line in space at each moment of time. All interactions in string theory take place by the splitting and joining of strings, with all particles arising as different vibrations of the same elementary string. It is the way a string vibrates that determines a particle's mass, its electric charge, its spin and other properties. This suggests there is no fundamental difference between particles of matter and particles of force, with matter and force different aspects of the same fundamental energy, aka $E=mc^2$.

String theory is the only theory that has been able to mathematically combine and describe the known forces of gravity, electromagnetism and the two nuclear forces into one set of equations. Accordingly, strings have been described as the threads that hold space-time together.

There is no need to get into the complexities of string, super-string and M-theories here, other than to note that they suggest the universe should have 10 dimensions of space and one time dimension. We are unaware of these additional dimensions because they are far too small for us to notice. We see only one time and three space dimensions, in which space-time is fairly flat. It is like the surface of an orange, if you look at it close up, it is all curved and wrinkled, but if you look at it from a distance, you don't see the bumps and it appears to be smooth.[10] (Also consider riding up in a very smooth elevator, if the floor signal light is not working, you might not even know you are traveling through the third dimension of height.) Similarly, it could be argued that electromagnetism, strong and weak nuclear forces and gravity could represent four of these other dimensions and that we do not notice them as we are so used to them.

What happens when 11 different dimensions interact, is this where things "spiritual" reside? Research suggests not, rather the result is something called a zero-brane. With zero-branes conventional notions of space and of distance between points melts away, according to Greene.[11] This opens up all sorts of possibilities about interconnectedness.

This interconnectedness is gradually being better understood by scientists, especially recent work in what is called entanglement. This is where one particle can instantaneously influence another particle (after previously being entangled with it), even if the other particle is at the other end of the universe. Researchers are just at the tip of the iceberg in their investigations into this intriguing field and it is as yet uncertain if the quantum physics of entanglement play any parts in our electromagnetic waves and circuits.[12] Accordingly, new age practitioners should not rush to use this as a scientific rational for their practices, as has been the case with some other discoveries of quantum physics: more research is required.

While research into string theory and this interconnectedness has a long way to go before we can understand what, and if any, role it has to do with the soul and spirituality there are many suggestions from antiquity that this underlying string energy has some role.

Verses of the Judaic *Kabbalah,* for example, sound very much like an ancient description of string theory and universal energy flows. "I am mustard seed in the middle of the sphere of the moon, which itself is a mustard seed within the next sphere. So it is with all the spheres—one inside the other—and all of them are a mustard see within the further expanses. Your awe is invigorated, the love in your soul expands." It also notes "the essence of divinity is found in every single thing—nothing but it exists. Since it causes everything to be, no thing can live by anything else. It enlivens them; its existence exists in each existent." This is echoed in: "Since it transcends and conceals itself, it is the essence of everything hidden and revealed."

Similar, the Chinese *Tao Te Ching* records "the way gave birth to unity, Unity gave birth to duality, Duality gave birth to trinity, trinity gave birth to the myriad creatures. The myriad creatures bear yin on their backs and embrace yang in their bosoms. They neutralize these vapors and thereby achieve harmony." The *Tao Te Ching* also refers to this universal energy as, "gossamer it is, seemingly insubstantial, yet never consumed through use".

There are also some interesting writings about energy (and possibly dimensions) in the Ein Sof section of the *Kabbalah*. "Emanating from the Ein Sof are the 10 Sefirot. They constitute the process by which all things come into being and pass away. They energize every existent thing that can be quantified. Since all things come into being by means of the Sefirot, they differ from one another; yet they all derive from one root. Everything is from Ein Sof; there is nothing outside of it. One should avoid fashioning metaphors regarding Ein Sof, but in order to help you understand, you can compare Ein Sof to a candle from which hundreds of millions of other candles are kindled. Though some shine brighter than others, compared to the first light they are all the same, all deriving from that one source. The first light and all the others are, in effect, incomparable. Nor can their priority compare with its, for it surpasses theirs; their energy emanates from it. No change takes place in it—the energy of emanation simply manifests through differentiation. Ein Sof cannot be conceived, certainly not expressed, though it is intimated in every thing, for there is nothing outside of it. No letter, no name, no writing, no thing can confine it. The witness testifying in writing that there is nothing outside of it is: 'I am that I am'. Ein Sof has no will, no intention, no desire, no thought, no speech, no action—yet there is nothing outside of it."

Then there is the *Bible's* more basic: "God said: 'Let there be light!' and there was light. God saw how good the light was and God separated the light from the darkness."

These and many others scriptures, coupled with the discoveries of quantum physics, suggests that we are still on the right path in terms of investigating energy as a major component of the soul and spirituality.

In fact, it could be argued that quantum physicists have recently made great strides in quantifying what these ancient writings attempted to describe. There is just one problem, string theory is purely mathematical; no one has devised an experiment to verify it. Also, while these ancient religious texts appear to be describing the same fundamental energy force as strings, we have to be careful in taking such writings than anything more than ancient peoples' recognition that there was more to life than they could easily see.

None-the-less, it is interesting that most religious traditions see life and the soul as emanating from some form of vital energy.

Has string theory revealed some of the essence of the creator in the underlying energy that creates us and the world around us? An energy-based force would encompass the minutest detail of the universe to its greatest grandeur.

Electromagnetic, nuclear and gravitic forces are indeed fundamental to the universe and life would not exist without any of them.

Even the bonds these forces form can only be broken by even further energy being applied. Accordingly, any supreme power, force or unified field would comprise an astronomical amount of energy, something several ancient texts suggest.

Interestingly, the scientific laws of thermodynamics state that energy cannot be destroyed, allowing for an energy-based creator to be infinite—and maybe even aspects of the soul.

For our investigation, the physics of waves of energy objectively explain the fundamental principles and mechanism of not only the mind, but also its elevated elements such as the soul and how it expresses spirituality. Remember how coherently resonating waves make our nervous systems work more effectively and provide a range of physical and mental benefits, such as increased immunity and increased consciousness.

Then there is the very important element of stochastic resonance. The physics of stochastic resonance involves a random background or stochastic fluctation or "noise" of electric waves. These noise waves add to the original waves, amplifying parts of them and/or decreasing other parts, as per the principles of wave interaction.

When additional waves are added in just the right manner, a positive stochastic resonance can be invoked, amplifying electric waves even further and heightening our senses, allowing us to perceive things we normally would not be aware of. This is a little like tuning a radio dial to just the right frequency to get a better reception.

For very weak or previously undetected waves, this stochastic resonance can result in amplifying the original wave pattern to a threshold where it can be detected. This is one way crayfish detect weak vibrations of potential prey in the water around them.[13]

Scientists have also discovered that random energy vibrations can increase the ability of nerves and the nervous system to recognize previously undetectable signals.[14]

These nerves include those chattering neurons that we encountered earlier, and that are responsible for LTP.[15] More importantly for us, this resonance appears to be involved in human consciousness. Stochastic resonance provides a mechanism for how something, such as a particular unconscious perception can be amplified above the threshold of consciousness and make it readily apparent to us,[16] a true extra sensory perception (ESP).

Besides heightening consciousness, stochastic resonance is also being used to improve the sense of touch in some people and balance in others.[17] And it could even be used in other biomedical applications to compensate for various brain dysfunctions, suggests some scientists.[18]

The physics suggests that stochastic resonance could also raise electric currents to the level of a spiritual experience.

This could explain why and how people who report spiritual experiences recall sensing a force inside and/or around them that they were not able to sense prior to such an experiences. The stochastic resonance allows them to be able to perceive what is there, but what was previously unrecognizable. Interestingly, the scriptures of various religions suggest that God is all around us, but that we are unable to notice this until we become spiritually aware.

There is also another form of wave motion that could be involved in generating LTP, stochastic resonance and maybe spiritual experiences. In pure liquids, such as water, individual molecules are in constant motion in every possible direction in what is commonly referred to as "Brownian motion". In human tissues, the presence of various larger elements, such as molecules, cell walls, organs and so on has the effect of dampening this motion. In some very structured tissues, such as the cerebral white matter of the brain and nervous tissue, there is a preferred direction of motion for energy. The movement of water molecules is restricted to directions that are perpendicular to the longitudinal axis of myelinated axons in the brain's white matter. This is above the direction of energy flow created by neurons. More simply put, this random Brownian motion is directed and ratcheted-up, making those electric currents stronger than before—and again able to trigger a threshold that they were not able to before. The end result is similar to stochastic resonance in that the nerve circuits become more sensitive and able to perceive things that they could not before.

Other experiments have shown that a fluctuating electric field can also enhance the flow of ions across cell membranes, the flow that creates the initial immaterial energy within ourselves. Dean Astumian reports that energy from the oscillating electric field substitutes for the energy provided chemically by adenosine triphosphate or ATP.[19]

There is another important wave motion, acceleration. So far we have tended to deal with waves of constant motion; yet when waves accelerate they can also bump into waves in front of them, creating stronger ones, such as the shock waves of explosions. It is uncertain if wave acceleration is enjoined by nerves and the body. The Institute of HeartMath notes the way heart waves vary is extremely important. And acceleration and deceleration is obviously one way that waves can vary. Quantum physicists also note that acceleration changes many equations regarding energy and provides different outputs than when on object is undergoing constant motion. Further research is required in this area.

The next question is just what are these sensitized nerves, minds and spirit sensing that they could not perceive before?

14—Part of Something Bigger?

Just what are we sensing with our heightened spiritual perceptions?

Spiritual texts provide a clue, with many of them suggesting our souls can sense energy from outside the body, from the universe.

The popular concept of "the Force" as portrayed in the Star Wars movies, was not just a stroke of luck for Star Wars developer and director, George Lucas. Lucas, together with famous mythologist Joseph Campbell, identified the concept of "the Force" as a myth common to many of the worlds' cultures.

Several spiritual, particularly Eastern, texts suggest that a human connection can be made or found with the universe or a universal energy—or what some people describe as becoming a part of a greater universal whole. Chinese spirituality, for example, has for thousands of years referred to people being able to absorb the life force Chi from the environment or from the universe.

The Chinese *Tao*, in particular, claims there are various heavenly forces that either provide us with energy or directly affect our lives. For instance, Taoists say "there is life-force in the air, but it is separate from the major elements of oxygen, nitrogen and carbon dioxide. The air is also made up of ions (charged protons and electrons), which are often depleted by pollution, overcrowded conditions in our cities, air conditioning, concrete buildings and other features of modern society. Without these energy particles, people become weak and tired. Science is studying ways to replaces these particles in areas depleted by human tampering, but to absorb their energies, we must separate them from the air, just as fish separate oxygen from the water. Taoist masters use the mind, the inner eye and the heart to distinguish and categorize the various forms of chi in nature and the universe," says Mantak Chia in *Awaken Healing Light of the Tao.*[1] He suggests, "Breathing is one of the most important functions that sustains our lives. What most of us don't realize, however, the cosmic particles of energy in our atmosphere as just as important to our survival as oxygen. (These ionized particles are also referred to as the Later Heavenly Force). Charged electrons supply the electrical current within our cells. If these are depleted, we can become weak, tired, and depressed, or we might suffer from negative emotional states and even physical illness. Although these substances are primarily introduced into the body through food and drink, they interact with and are affected by the ions in the atmosphere. The earth's charge is positive. To balance this and maintain optimal human performance, we need negative ions. There are abundant negative ions in the air in the mountains and forests, by oceans, rivers and waterfalls, after a rainstorm, and in grottoes. This is why sages of all traditions usually seek to live close to nature. Taoist meditations help still the mind and body so we can absorb, refine and conserve higher energies," say Chia and the Universal Tao Center.

Can modern science shine any enlightment on this apparent scientific impossibility of absorbing energy directly from the world around us?

We already know that ionized particles are a fundamental of life, with the ions in our cells providing the energy for movement, thought and more. Normally these ions are made from the food and fluids we consume.

It is also known that a change in the type and number of ions in the air, caused by weather such as storms, can also affect is, in particular our emotions. A predominance of positive ions or lack of negative ions has been found to be correlated with changes in the level of serotonin. Too much serotonin affects appetite and sleep and too little affects the mood, often causing depression. (Just how this natural chemical works and its relationship with ions is still being explored.)

In contrast, an abundance of negative ions can help excite neurons, helping you feel sharp and your brain to focus. Lightning puts ozone into the air, and ozone has been shown to help people feel slightly euphoric, as have negative ions at the beach or in the mountains. If you sit on mountaintop or by the ocean you may breathe in negative ions of oxygen.

Research has also shown that the amount of sunlight we receive can increase the amount of serotonin in our bodies. While it undeniable that our bodies absorb sunlight, but we are not yet certain of the full energy transfer process here. We know we are not quite like reptiles that need sunlight to warm their bodies and blood, though this does demonstrate the power of the radiation contained in the sun's waves of light.

There is also other evidence that other "heavenly forces" affect us. For instance, there is also the gravity of the moon, which affects the tides and other monthly cycles, as well as some electric potentials on life, as mentioned earlier. Also, it is known that such celestial electromagnetic events as sunspot cycles, eclipses and magnetic storms in the earth's ionosphere also affect people. Japanese researcher Maki Takata found that the precipitation of the protein albumin in blood varies with these events as well as the date of the year.[2]

While, scientists readily accept that sunspot activity and magnetic storms have substantial affects on radio and other communications (proof that cosmic energies affect energy wavelengths) many are still less inclined to accept the evidence that these things also influence people. In contrast, astrologists claim these heavenly influences account for our personality, according to when we are born.

There are also other types of energy flying through space with the earth bombarded by cosmic particles such as photons, electrons, protons and heavier particles. Some of these particles have been measured as having the energy of a tennis ball traveling at 100 miles an hour, all concentrated into something smaller than an atom. It is uncertain what affect, if any, the have on us.

This leads to the question is space just an empty vacuum or is it filled with something else, a universal energy, an energy that might impact us? Many ancient cultures believed the latter to be the case. So did Isaac Newton, who noted that a body spinning in a vacuum, such as a rotating planet, experienced a centrifugal force, (the earth bulges slightly, just below the equator as a result of this). Newton thought there must be something in space pushing back on the spinning body to create this force. Einstein thought the gravitational field of the rest of the universe might explain centrifugal and other forces of inertia resulting from acceleration. However, when he finished formulating his general theory of relativity in 1915, he was disappointed to find that it did not seem to incorporate this.

Two 19th century scientists, Michael Faraday and James Maxwell believed magnetic and electric fields were stresses in a universal medium, which became known as the luminiferous ether. Maxwell believed electromagnetic waves, such as light, were vibrations in the ether.

The idea that we are surrounded by this ether appealed to spiritualists of the day, who believed that we had an etheric body as well as a material one, and the tradition has continued unfounded among some new age practitioners.

Today, recent scientific measurements indicate that space does indeed contain an ever-present energy flowing throughout the universe. In fact, cosmologists believe the majority of the material that makes up the universe is "dark energy". Our 13.89 billion year old universe comprises just 4% of the type of matter that we experience in our everyday lives, with more mysterious "dark matter" accounting for another 23% and the remaining 73% believed to be "dark energy".

Scientists are unsure what comprises 96% of the universe we inhabit.

This dark energy is believed to be in the form of as yet unidentified fields of force. Some cosmologists suggest it is an "anti-gravity" that pushes the universe apart called quintessence.

This energy was first proposed by Albert Einstein and he labelled it a "cosmological constant," but later dismissed it as he estimated its value to be zero.

Interestingly, in another study, Davies found that acceleration of an object in a vacuum creates photons. He found that by accelerating two opposing mirrors in a vacuum that he was able to create photons where none had been before.[3] Is this due to the light between the two light mirrors being distorted to the point where new photons need to be created to convey it?

Whatever the case, this sounds like something being created out of virtually nothing, something akin to an act of God—an energetic God. However, some scientists have since suggested that what Davis found was the generation of energy waves, waves that build upon each other to ultimately build elementary particles, such as photons.

Another theory about the ether is that it is made of light (photons, as well as virtual photons) and this is what pushes space apart. Physicists Bernard Haisch and Alfonso Rueda suggest that photons in the quantum vacuum resonate at a specific frequency. Remember how the speed of light is constant for all observers, no matter how fast they are traveling toward or away from the source of that light, in contrast to slower radio and other waves. Einstein resolved this paradox with his theory of special relativity and demonstrated that the flow of time is inextricably connected with space—or the field. According to special relativity, no longer can space and time be thought of as universal concepts set in stone, experienced identically by everyone, but rather experienced individually.

Only further research will provide a better indication of just what the space of the universe contains, and what dark energy might comprise.

Is it the radiation of dark matter, dark energy or the affects of one of those other dimensions identified by string theory that people are sensing when they report spiritual experiences that make them "one with the universe" or a greater power? Just what is being sensed through the heightened perceptions created by stochastic resonance?

Using the radio or television wave analogy, it appears that we are sensing something, but are just not sure what is being transmitted, how it flows, and certainly not quite what it means. Though, by many accounts it appears to have great meaning to us. We just have to decode the message.

Again, there clues from spiritual texts that provide a possible way forward in terms of our relationship to external energy.

Consider the ancient Rishi sages of India who said it is energy and time that orders the universe. The Rishis postulated five energy fields, with the fifth being the field from which all things emerge. The threshold of this final field is where time and space, spirit and matter intersect and begin to evolve. They also described a state of 'absolute being' in which there is a fusion of time and consciousness, a timelessness where time-energy is said to vibrate at such a rate that it appears static. They believed the difference between people and the universe was a matter of degree, rather than composition. If a person could align themselves with the harmonies of cosmic energy and time they would gradually become aligned with the cosmos and creation force.

Many Eastern religions similarly consider that God is everywhere and that it is up to us to realize this and try to become one with this "grand organizing designer".

The Dalai Lama, Buddhism's contemporary leader, suggests that we do not have to develop our souls but simply become part of the universe.[4]

Some scientists have suggested a link between man, energy and spirituality. Even Einstein once said there must be something behind the energy.[5]

Hunt suggests, "The energy field is the highest level of the mind of man."

She says "divine vibrations" are available constantly and contain all the information we need. Human perceptions are the weak link.

For instance, she notes, "In my laboratory, we found that when a person's energy field reach the highest, most complex [energy] vibrations, from imaging or meditation, that person had spiritual experiences, regardless of their beliefs. Even though the imager was culturally-linked, each person identified the experience with a divine essense that was beyond any specific religious belief system."

She adds, "I concur with the statement of contemporary theologians, 'God is indeed within man.' But I do not believe that man can perceive his 'God-like' qualities until his [energy] field reaches higher vibrations and attains a greater degree of coherency. No matter how hard we try to receive spiritual guidance, we cannot until our fields are attuned to that vibrational system."

In contrast, physicist John Wheeler proposes that human actions have an impact on the universe. He theorizes that many things are possible in terms of quantum physics, but these possibilities do not become real until they are observed. Our observations might contribute to the creation of physical reality, he suggests. As such, Wheeler conjectures the universe is a work in progress and that our interactions with it create and shape the world in which we participate. The universe is an enormous feedback loop, he speculates. This is not just a theory, as experiement at the University of Maryland in 1984 suggested that the way we measure things, such as photons, determines what is real and what is not.[6]

Bohm suggested something similar, in that quantum waves carry information, and that this is potentially active everywhere, but is actually only active when it can be decoded, such as when it interacts with an appropriate particle. For example, radio waves can travel throughout space but can only be decoded by the electromagnetic energy and components of a radio receiver.[7]

Is spirituality information that we have not learnt how to fully decode yet?

These concepts of a human-universe connection date from antiquity and are common across cultures and countries—and refer to something that today sounds a lot like the interaction of energy fields.

15—The Force of Fields

Today's scientific study of energy fields is providing some fascinating findings, revelations that could shed light on our souls and spirituality.

Most particles, such as electrons and quarks, have an oscillating electric charge or 'spin' and behave as tiny magnets, with the direction of each magnet aligned along the direction of the spin. While a single spin generates electromagnetic waves, several spinning particles can produce a small electromagnetic field.

In turn, the individual charged particles interact with the overall electromagnetic field that they have produced. Charges produce electric and magnetic fields and these fields in turn exert forces on the charges.[1] Interestingly, photons are electrically neutral and do not produce an electromagnetic field.

This point is important in that while the particles create the field, the field in turn influences the action of the individual particles! For example, particles create gravity and without them there is no expression of gravity, yet gravity also influences and even directs those very same particles. One way fields interact with each other is via waves.

In a more everyday example, consider how ice cubes melt in a glass of warm water, yet do not in cold water. Consider an egg in a pot of cold water, there is little change; yet when the water is boiling, the egg inside its shell becomes firm. In these instances, there is no chemical exchange of atoms; the change is the result of an electromagnetic change in the surrounding field. And there are many other examples in everyday life where the field, be it air pressure or something else as "invisible", can have a great effect over a wide region and many people and particles. The field also appears to be involved in what comprises the difference between an assemblage of atoms and a living object, such as a plant. The interaction between electromagnetic and biochemical fields is also integral to consciousness.

In an electromagnetic field, charged and neutral particles do not move about the same way. It appears that charged ones somehow know something about the field that neutral ones do not. John Taylor in *Hidden Unity in Nature's Laws* suggests this is due to interaction in other dimensions that we can not see,[2] while other say it is due quantum electrodynamics where electrons give off and absorb photons that seemingly "sense" what is around the electron.

The importance of fields was noted by Einstein, who described matter "as being constituted by the regions of space in which the field is extremely intense".

He said, "There is no place in this new kind of physics for the field and matter, for the field is the only reality."[3]

Similarly, Burr said, "The universe in which we find ourselves and from which we can not be separated is a place of Law and Order. It is not an accident, nor chaos. It is organized and maintained by an electro-dynamic field capable of determining the position and movement of all charged particles. For nearly half a century the logical consequences of this theory have been subjected to rigorously controlled experimental conditions and met with no contradictions," he said.[4]

More recently, Haisch, Rueda and others suggest it is the interaction of charged particles and the electromagnetic field that creates the appearance of mass. The book you now hold in your hands is "massless," it is physically nothing more than a collection of electric charges embedded in a universal electromagnetic field and acted on by the field in such a way as to make you think it has the property of mass. Its apparent weight and solidity arise from the interactions of charges and the field, Haisch and Rueda say. "The charge is the source of an electric field, which carries energy—the electrostatic self-energy. A basic result from classical electrodynamics is that a fluctuating electric charge emits an electromagnetic radiation field. The result is that all charges in the universe will emit secondary electromagnetic fields in response to their interactions with the primary field," they note.[5]

This is a little like the interaction of waves mentioned earlier, but in more dimensions, creating new waves and fields.—and in turn more interactions.[6]

These, and other findings, suggest that energy fields (be they electromagnetic, gravity, dark energy or other fields) are the fundamental basis of, and the overriding governor, of everything. They can be huge, as in gravity fields, or minute as in the field of an electron and its photons rotating around the nucleus of an atom. The late physicist Richard Feynman believed the movement of electrons and photons were two of the three fundamentals of the universe, the third being the fact that an electron emits of absorbs a photon.

There may also be a field form of stochastic resonance, heightening the interaction of some fields upon others that would not normally be noticeable.

Also, this interaction of one field on another can be very close or very distant. Davies suggests the quantum ether provides a form of connectivity, a connection between local and distant effects. Quantum physics is famed for its 'no-locality,' the fact that it is not possible to characterize the physical situation at a point in space without reference to the state of the system in the wider surroundings, he notes. The quantum vacuum is no exception, since its state is defined across all of space. This enables it to 'feel' the structure of the entire universe and thereby to link the global and the local together.[7]

Bell's theorem also suggests that all objects and events are inter-connected with one another, and has been further expounded upon by the theory of entanglement. This is where particles that have become quantumly entangled maintain a relationship, influencing one another, even though they may be great distances apart.[8] It is as yet uncertain if entanglement works on a wave, field or other quantum level.

However, it does provide evidence that some of the atomic particles inside us may be connected with other particles located outside of our bodies, even in distant locations. Does this have an impact on our souls and could it account for why we sometimes feel a connection with the greater space or universe around us? Only further research will tell.

Then there is James Lovelock's ecological concept of Gaia, of an interconnectedness of not only everything on the planet, but also in the universe.

Bohm suggested energy fields are arranged in a vibrational hierarchy, representing energy in successive states of manifestation from infinitely subtle states to those that are seen in the physical world around us and appear solid.[9]

A key question is accordingly what comprises a field? At its most fundamental level we know that this is "strings" of vibrational energy. Whether this function at the individual string level or are combined into something else such as dark energy or photons remains to be determined.

Whatever it comprises, it will also be influenced by others factors adding their influences here and there. Fields upon fields, like layers upon layers (such as gravity upon electromagnetic field upon other fields) is what gives us the universe, the earth and ourselves such complexity and individuality.

Bohm, a protégé of Einstein, suggested the field is also accompanied by guiding waves that permeate space-time. The intensity of these waves does not diminish with distance and they exert no force on particles. Essentially, they are waves of unifying, undecaying information that cannot be broken down into either a wave or a particle, but are always present as both a wave and a particle.

This could help explain the interesting phenomenon of how images of ourselves are right now voyaging through space. Similar to how we can see other stars in the universe, someone with a powerful enough telescope in another galaxy could look at our little planet and see you reading this very word, right now, even though the moment has passed. The photons of light reflected from your own body are being radiated into space (if you are outdoors or near a window) so that someone far away could see you in several light years, long after your body has disappeared. So, in a sense, your image remains traveling at the speed of light throughout the universe—a kind of visual immortality.

Interestingly, Bohm notes that as something moves faster and faster, its time rate slows down according to relativity, so that as it approaches the speed of light its own 'internal' time becomes less, virtually standing still—a timeless state. In other words, light has a timeless quality or as Bohm said, "time originates out of timelessness". Does the passage and interaction of the photons of light create time, relatively speaking? While physicists continue to research photons and their importance, many scriptures have been referring to light, time and their interaction and relevance to life since antiquity. God's reported remark, "let their be light," as 'he' created human life being one of the most famous references.[10]

In contrast, Bohm suggested, "Light is what enfolds all of the universe as well. For example, if you're looking at this room, the whole room is enfolded into the light which enters the pupil of your eye and unfolds into the image and into your brain. Light in its generalized sense (not just ordinary light) is the means by which the entire universe unfolds into itself.…At least as far as physics is concerned." He noted that light "is energy, and it's information—content, form and structure. It's the potential of everything." He suggested that light, "by interaction of different rays (as field theory physics is investigating today), can produce particles and all the diverse structures of matter."[11]

In terms of our search, photons, electrons and electromagnetic fields are also known to influence the very fundamentals of ours bodies and minds. The flow of ions into and out of nerve cells, for example, creates one form of electromagnetic field, the perineural cells around nerves another—together providing the spark for LTP and consciousness as we noted earlier. These fields, along with other fields created by brain waves, heart waves and so on creates a bodily field.

In neurons, certain electric fields are known to be able to cause ionic charges to build up and induce ionic currents both inside and outside the nerve cells. The field changes the ionic potential at the nerve cell membrane, shifting the firing threshold for nerve cell action potentials. When the electric field lowers the threshold, a form of stochastic resonance can apply, where nerve cells fire at a level that would normally be insufficient to trigger them into signaling the next nerve.[12] This electric field could be generated by perineural currents or some other influence. Whatever the case, here we again encounter the importance of stochastic resonance, this time in terms of electric fields.

Some external fields obviously influence our own bodily field to varying degrees. This can range from short intense-energy fields to more subtle low-energy fields that have longer-term affects. For example, while many external fields have no or little impact on us, the concentrated force of the field that accompanies a bullet or an electric shock can kill us. The microscopic, yet more widespread fields created by viruses can also be fatal, while in other cases they do not. It is how these forces and fields interact with our own that is key.

Our own electromagnetic field can also influence certain fields around us, which in turn can react and reinfluence our field. This could explain how the observer influences the field, as detailed by Schrödinger and his infamous cat paradox.[13]

Researchers have discovered that proteins in our cells "sense" electric fields and are influenced by them, changing shape to different energy frequencies.[15] This may occur via the 18,000 or so electrons that are thought to be inside each protein. (Interestingly, if these loosely bound electrons are accelerated they will emit photons in the infrared wave range.)

This changing of shape by proteins is involved in a myriad of processes inside cells, such as ion channel opening and closing, as well as how they interact with genes.

This is where the quantum world controls chemistry and biology.

New research also shows that we can influence our own fields, with research by Jeffrey Schwartz, Sharon Begley and others showing how the mind possesses a plasticity that enables it to influence itself and the body.[14] (More on this later in Part III.)

Enzymes are known to be able take energy from an electric field to speed up reactions. Interestingly, this action is also frequency specific, in particular, low-frequency specific (around 10Hz) according to James Weaver and Dean Astumian.[16] They found that cell membranes use energy from electric fields to perform certain functions. "Cells use dynamic electric fields to accomplish the necessary functioning of life." They also note that proteins change shape back and forth, with different shapes sensitive to different electric fields. Proteins store free energy by absorbing electrical energy when they change shape. "The energy absorbed from the electric field is not just dissipated as heat. An enzyme, for instance, can use it to accomplish valuable functions such as transporting nutrients into the cell against a concentration gradient, forming ATP for the cell to store for later use, and signaling to the inside what is going on outside."[17]

Another study found that electromagnetic fields influence how cells relay directions to DNA.[18]

And vice versa, with another study finding that DNA can also vibrate and send signals throughout the body to energize proteins, which in turn trigger chemical reactions and so on to ultimately produce life as we know it.[19]

A further study found that electromagnetic fields can also directly affect our DNA and switch on genes in damaged nerves.[20]

In fact, it now appears that energy is a primary factor involved in the initiation and signalling of DNA to replicate cells to ultimately create ourselves and our offspring.

This interaction of energy and DNA would seem to meet the definitions of the soul as an underlying principal of life, and the incorporeal or non-material part of our selves.

Hunt found an individual's energy field changes before specific waves flow, such as brain waves, blood pressure, galvanic skin responses, heartbeat and muscle contraction.[21] This was also found to be the case in experiments by Benjamin Libet, who noted brain activity occurred before conscious recognition of a particular thought. Hunt also suggests that each person has a unique electromagnetic field.

This theory of energy fields influencing and directing our minds and bodies was taken a step further by Burr who theorized "the pattern of organization of any biological system is established by an electrical field which is at the same time determined by its components and determines the orientation of the components. The field maintains the pattern in the midst of a flux of components. This is the mechanism whose outcomes is wholeness, organization, continuity," he wrote in *Blueprint for Immortality.*[22]

Burr developed a concept of "life fields", or L-fields as he called them, which "impose design and organization on the constantly changing material components of all living forms. They compel an acorn to grow into an oak tree—and only an oak tree; they compel a maize seed to grow into a corn stalk and not a stalk of wheat or barley. Anything that compels growth and development in an organized way is irrefutable evidence of law and order. But the L-fields of this planet are themselves influenced by the greater fields in which our world is enmeshed, as we found from the effects of sunspot activity on the fields of trees. They are subject, then, to 'higher authority'—so to speak, which compels them to change in various ways. And, no doubt, the fields that surround this planet are themselves subject to the greater fields of space. In other words, L-fields are links in a 'chain of authority'. This starts with the simplest living forms, runs upwards through all the life on this planet to the most complex form we know—man—and then extends outwards into space and upwards to an infinite, ultimate authority, about which we can only speculate. This 'chain of authority', of course, does not apply only to the fields of life. It must also extend from the heart of the smallest atom to the gigantic forces which keep the planets in their orbits, which govern the stars in their courses and which regulate the feverish race of the most distant galaxies towards the outer reaches of space." Unfortunately Burr did not elucidate on the physics of just how this happened.

Interestingly, it also appears that recalling and invoking some memories can result in the replaying of the energy pattern associated with that memory, (much like the reigniting the association of the molecules of emotion of the memory). For example, memories and thoughts of love can cause the heart to beat faster and waves associated with this feeling to be conveyed throughout the body; or in the case of a fear, more jagged waves of energy can cause your skin to sweat. This is interesting in that while the initial stimulus may no longer be there, the mind can cause the body to undergo a similar wave and field response as if it were still present.

Scientists continue to investigate such fields and how they may function—either through the interaction of photons (waves and particles of light), gravitons (unproven particles of gravity) or through the little understood dark energy we encountered earlier.

For our purpose of investigating the soul and spirituality, these discoveries show that energy fields can function at both the immense scale of the universe as well as at the microscopic level as within our nerves, proteins, cells and DNA. This equates to many of those ancient and religious definitions of a spiritual force that can encompass and comprise the universe as well as ourselves.

There is also another type of field that provides a framework of how energy fields could work within our minds, and possibly our souls.

Holographic fields, or holograms, are created by two beams of tightly focused energy or coherent laser light intersecting each other on a photoreflective substance. Where the beams intersect creates a pattern of light and dark, with the photoreflective substance shifting electrons to the dark or negative region of the pattern. These microscopic patterns of light and dark (not just on and off) are what ultimately create the holographic image.

In the past, holograms were traditionally created by just one color of laser light. Using a wider range of light provides the ability to store huge amounts of information. Scientists have verified this; by varying the wavelength/color of the laser beam or the angle the lasers intersect on the recording medium they can store different holograms in the same location.[23]

A holographic field could be a crude analogy as to what happens when LTP occurs in our minds and how LTP stores information in memory. Coherent waves of electric nerve and/or perineural current would replace the laser light.

Now imagine this as a dynamic process, where images and memories can be refined, updated and even altered. This could also explain how we not only see 3D physical images in our mind, but also images over time—effectively 4-dimensional movies as well as concepts. Then there are all the words, the language and other bits of information that do not require images, even more of them can be stored in the circuits of our head. (We won't delve into the elaborate chemical actions and reactions that ensue, as while they are very important to the process and life in general, they do not appear to be the primary or elevated element that we are seeking for in our quest into the workings of the soul and spirituality.)

This holographic field theory of the mind and soul is well beyond the traditional descriptions of the brain as some form of computer, as the closest we have come to a field computer is parallel processing by several computers at once, which still falls short of what the energy in our brains can do. The brain could using the two electric systems noted above to create a sort of parallel processing, with emotions possibly adding a third system to provide greater depth of processing.

However, there have been recent advances in computer storage involving holography that provide an explanation that could account for the complexity of our brains. In holography, data is projected from two or multiple sets of laser beams onto a recording medium such as a crystal. Altering the angle and/or wavelength of the beams mean that many images can be recorded in the same medium at the same location. To recover the data, a reference beam of light is shone through the crystal medium at the same angle used to record the page of information and strikes a camera that reconstructs the original image. This technique allows for a thousand-fold increase in storage in the same space as digital bit storage—and without moving parts.

Nobel laureate, Karl Pribram, sumizes "brain research tells me the brain is working along holographic principles".[24]

Such a "holographic mind field" could also account for how the atoms and cells in our body are replaced, yet we retain a coherent recollection of ourselves and others. The body and its atoms can change but the information continues as this holographic mind and "soul" is not necessarily a physical part of the body, but rather electromagnetic circuits and information, which are developed and housed in the body. (Do they continue after the body has gone? That's another story.)

For example, just as a little piece of our DNA can store all the information required to reproduce our bodies, so too can a hologram. Cut off a small piece of a hologram and you can recreate the whole image from that small snippet.

This could also explain when part of the brain is damaged or removed that it generally merely "dims" the operations of the mind and certain memories, rather than or erasing total brain functioning. Neurobiologist, Karl Lashley tried to determine where memory was located in the brain of rats. First, he trained his rats to find their way from one end of a maze to another, then he systematically removed small amounts of their brains, one area at a time. He reasoned that at some stage he would remove a piece and they would no longer remember how to run the maze. However, after removing some 90% of their brains, the rats still remembered how to find their way through the maze, with only a slight decline in accuracy and speed.

As for emotions, they would appear to add another dimension to the holographic information that is recorded and stored. Remember how the brain adds emotions to thoughts to indicate their importance. And the emotions added to spiritual experiences suggest they are extremely important to us.

It could also explain how we each see the world and life a little differently to each other, the way we record images as being slightly different to each other.

Bohm, for example, theorized the way we see things is often like looking at a fish swimming in a fish bowl. Two cameras film the tank from different directions. We see only the two films project on two screens, two dimensions, two sets of light waves. We think we are watching two fish, with some curious correlation in the way they interact. The bowl of water is in an unseen higher dimension. However, the two fish we think we see are really projections of a single entity in a unified world.

He suggested there is also a guided wave in the bowl, or in a hologram, that enfolds and conveys information about the entire picture. Bohm suggested an example of enfoldment was what takes place when a person listens to music. "At a given moment, a certain note is being played, but a number of the previous notes are still 'reverberating' in consciousness. Close attention will show that is the simultaneous presence and activity of all these related reverberations that is responsible for the direct and immediately felt sense of movement, flow and continuity, as well as for the apprehension of the general meaning of the music. To hear a set of notes so far apart in time that there is no consciousness of such reverberation will destroy altogether the sense of a whole unbroken living movement that fives meaning and force to what is being heard. The sense of order in the above experience is very similar to what is implied in our model of a particle as a sequence of successive incoming and outgoing waves."[23] This appears to be an extension of wave interaction and stochastic resonance, with an additional variable component of time.

This holographic field theory provides a crude understanding of how immaterial electromagnetic currents and fields could be responsible for the way the elevated elements of our mind may work and, which when coupled with stochastic resonance, could give rise to the elements of the human mind that are often attributed to our souls and spirituality.

It could also explain how we can, in a spiritual state, sense and understand the universe—in such states we might just perceive, decode and understand a piece of the holographic field.

Interestingly, holographic fields are also being used to explain the universe, with particles such as photons or dark energy believed to be holding and unifying these fields together as well as conveying waves of information throughout and influencing them.

Some theologians have also adopted aspects of field theory to help explain various aspects of religion. For example, W. Pannenberg suggests "the field concept in the development of modern physics has theological significance. This is suggested not only by its opposition to the tendency to reduce the concept of force to bodies or masses, but because field theories from Faraday to Albert Einstein claim priority for the whole over the parts. This is of theological significance because God has to be conceived as the unifying ground of the whole universe if God is to be conceived as creator and redeemer of the world. The field concept could be used in theology to make the effective presence of God in every single phenomenon intelligible."[25]

Burr took a slight different approach and suggested, if "the brain is the medium of expression of the human mind, similarly the entire physical universe would be the medium of expression of the mind of a natural God. In this context, God is the supreme holistic concept."[26]

16—The Power of The Soul

Just what power does the soul wield in our lives? Reports of spiritual experiences, from antiquity to today, suggest they can be incredibly powerful, with people changing their lives following such an experience.

As an elevated element of mind, the soul obviously has an impact on how we perceive, think, react and live. Some theologians suggest it is a moral compass that helps guide us. There are many religious metaphors as to how this operates. But, with each of the world's many religions offering different suggestions, I will avoid the controversies presented by a comparison of them in favor of more objective metaphors.

For instance, the nerve cells in our brains and bodies are a little like a band of musicians. There are a range of different instruments that each band plays win its own style that gives them distinct sounds. Together, the instruments, the way they are played and what they play comprises the music the band ultimately makes. The way nerve cells interact in terms of LTP resonance is a little like the instruments. The resultant brain waves are a little like the tunes being played, while the encompassing memories, experience, knowledge and even feelings we apply to the way we think are comparable to the arrangement of the tune. These electromagnetic flows are also enjoined by emotions and chemicals which add even further depth and make each perception, or performance, uniquely your own.

Another analogy is a company of people, each doing their job under an executive who directs them to pull together to accomplish a specific task. The way they get this task done is often referred to as "in the spirit of," such as they did it with a spirit of teamwork, or confrontation, or some other approach.

The band or company's approach and principles help guide them through whatever they do. An organisation that has instilled clearly defined principles tends to encourage employees with those principles to join its direction, and in turn further develop how the company operates. In contrast, organizations that don't have any such principles "float" along without any overall guidance, while those with "negative" principles (such as cheating, corruption, etc) often do not last long before they flounder.

In an individual, it is the spirit, principle or nature of perceptions, learning and memory that create the moral compass referred to above. A distorted "compass" could explain how some people can be religious, even faithful to the point of being "religious fanatics," but still be described as soul-less or evil. They act with a spirit of religiousity, but not the spirit of religion or morals that the majority of us appear to be using.

Another useful metaphor of how the soul works is that of a movie director. The way you interpret a story or script and how you develop this into a movie, compared to another director's, is based upon how you interpret that script, the emotions and thoughts that made you want to make a movie of it, why you want want to share the story with others, and your knowledge, experience and other cultural and personal references that you bring together to create the film.

This is also, in a sense, what we do when we percieve, learn and memorize things. We not only lay the building blocks of our minds, soul and spirituality, we create our own "holographic movie" in our minds.

Accordingly, if you have limited perceptions, knowledge and experience it is likely you will be restricted in the resources and creativity that you can bring to direct your own movie. How we construct our perceptions to create our holographic movies of what happens to us and our lives is a key determinant of our individual souls and spirituality.

While we can't change cultural references, personal experiences and what happens to us, we can change how we label what we percieve. How we label our perceptions plays a major role in how our minds and souls operate and direct us in life. For example, take what you consider good versus bad. The universe may have a range of opposing forces, such as exploding stars and collapsing black holes, but it does not distinguish between good and bad. There is nothing right or wrong about the way atoms interact. Yet virtually everyone knows instinctively what is right and what is wrong. How we distinguish and choose between good and bad/right and wrong is something our souls are considered to be responsible for.

Just how do you determine what is good versus bad? Traditionally, religion and community values have instilled this, along with education. Knowledge, experience and emotional intelligence also appear important.

There is also a physiological basis. For years, I wondered how people not exposed to religion or community values determined between right and wrong, good and bad.

While the electromagnetic energy flowing around our minds is impartial, there are "good" and "bad" brain, heart and other waves in terms of coherence.

Recall how negative thoughts and emotions result in jagged and erratic wave patterns, and how positive thoughts and emotions tend to generate smoother waves. These smoother waves are also more coherent, able to entrain other waves into their resonance, in turn strengthening it, while erratic waves tend to not.

This impacts LTP as it forms and references perceptions, learning and memory—and provides an account of how we could use impartial energy and code it as positive or negative, good or bad.

This would provide a mechanism whereby we "intuitively" know what is right and wrong, where we "feel" or sense that something is good or bad.

This suggests that we can also determine to some extent what resonates in our LTP, and therefore what we perceive, learn, memorize—and most importantly how we emotionally tag those LTPs.

Remember the work of the Institute of HeartMath and others which indicate that we can control our heart waves, and therefore to some extent our brain waves. This implies that we can influence not only our bodies with our thoughts, but also our minds and souls to some degree.

This does not mean that we can change ourselves into a completely different form, but scientific evidence certainly suggests we can influence the way we perceive and respond to situations. (Recall the power of suggestion, belief and hypnosis outlined earlier.)

Accordingly, the development of your soul depends to some degree upon how you modify your nerve circuits and synapses. Continued thoughts and feelings of positiveness and revelation would certainly encourage the re-use of certain synapses, as would continued fear (or indoctrination) reinforce other neural pathways. The latter might be easier, with several marketing studies showing that fear of loss is greater than desire.

Besides altering our nerves, other research has found the way we think can also effect our genes. Certain thoughts can cause the release of hormones that can bind to DNA, turning genes off or on, according to psychologist David Moore.[1] And we know that energy flows and fields can influence proteins and other cells.

Perception and emotional experience, in particular, are known to effect genes. For example, how much a mother rat interacts and licks her baby rats can determine whether genes that code receptors for stress hormones in the brain are activated and at what level in those babies. Baby rats with a mothering parent were much less fearful and more curious than those that received less interaction and who grew up to be timid and withdrawn.[2]

Interestingly, other research indicates that having a negative self perception can also reduce your life expectancy, while positive self perceptions can increase it. A positive attitude towards aging can extend a person's life by seven and a half years, according to researchers at Miami University. The way a person perceives aging outweighs gains in lifespan from quitting smoking or exercising regularly, found Suzanne Kunkel, director of the university's Scripps Gerontology Center. The study found respondents with more positive views on aging lived longer, even after taking into account factors such as socio-economic status, health and loneliness, living a median of seven and a half years longer than those who had negative perceptions. The will to live contribued to the relationship between positive self perceptions of aging and longevity, but did not completely account for the difference, found the study conducted over more than 20 years.

Again, it is perception and the associated electric processes of LTP that appears to make the difference, and that difference in perception comes from the elevated element of mind, our souls. This helps to explain why identical twins, with identical genes still have different personalities and spirituality.

Accordingly, it is not so much about controlling what happens to us, which often we can't, but rather controlling how we define it! How we channel and direct the electromagnetic energy flows appears to be a major power of the soul.

In summary, physics explains how a vibrational energy system, comprising nerve circuits and associated electromagnetic waves, can resonate to generate an electromagnetic field that perceives, learns, memorizes and resolves information.

This process provides an objective framework for operation of the higher level of our minds and outlines how our souls and spirituality function and direct various aspects of our lives.

When these waves resonate coherently in various parts of the body and incorporate a stochastic resonance they become even more sensitive and able to sense things previously undetectable or unconscious to us—or what we call spiritual experiences.

The greater the level of this functioning, the greater the sensitivity of awareness of our self and the world around us and of our potential. It is this awareness of our own mind, or our awareness of being aware, that separates our level of consciousness from that of animals. (Written communications are another distinguishing feature of the human level of consciousness).

However, what we have identified above is just the "hardware" aspects of the soul and spirituality, much like those inside a computer. These items sit idle until something or someone runs a software program on them.

The software in this instance includes the perceptions, learning, memories, experience and emotions that we store in our minds, and which also provide the references and "programs" that provide the basis of how we act and react.

Our very notion of personal identity—the self, the soul—is closely bound up with memory and enduring experience, says Davies author of *The Mind of God* and *God and the New Physics.*[3] "The essential ingredient of mind is information. It is the pattern inside the brain, not the brain itself that makes us what we are." He says, "The property of self-awareness is holistic and cannot be traced to specific electrochemical mechanism in the brain..... It is the very entanglement of the levels that makes you *you.*"

Davies adds, "This conclusion leaves open the question of whether the 'program' is re-run in another body at a later date (reincarnation), or in a system which we do not perceive as part of the physical universe (in Heaven), or whether it is merely 'stored' in some sense (limbo)?"

Yet, this common hardwired mechanism and individual software program that provides a framework for the soul and how we experience spirituality does not explain why we experience spirituality.

It is easy to understand why we experience fear, we would not survive long without it. But why do we need to experience spirituality?

While spiritual experiences tend to be rare occurences in most of our lives, there is another intense experience that many of us have, and whom many report as coming close to spiritual experiences—that of orgasm during love making. While these are mostly focused on the reproductive organs, they can also extend throughout the body and the mind. Tantic sex practitioners, for example, seek mental orgasm in conjunction with the more common physical ones.[4]

This raises an interesting philisophical question; if orgasm is an encouragement to procreate and reproduce a species, what is the even more intense spiritual experience designed to achieve? Is spirituality designed to ensure our species continues, or is it for something else, something which we have yet to discover?

The question of whether God created the soul and spiritual hard wiring for us to be capable of spiritual experiences or whether our brain wiring created God remains unanswered, with no obvious clues. What you believe is a matter of faith. Newberg and d'Aquili, for instance, suggest all religions originate from common experiences derived from the capabilities of the human brain. They say the brain does not invent religious states, but instead finds them. "A neurological approach suggests that God is not the product of a cognitive, deductive process, but was instead 'discovered' in a mystical or spiritual encounter made known to human consciousness through the transcendent machinery of the mind..... The accounts of the mystics are not indications of minds in disarray, but are the proper, predictable neurological result of a stable, coherent mind willing itself toward a higher spiritual plane....We believe all lasting myths gain their power through neurologically endorsed flashes of insight."[5]

What is at the bottom of your mind? God knows what.

The answer appears to involve some element of unification with, or striving to be closer to, God. I suggest there is an ultimate goal in spirituality and that we are in some ways too primitive spiritually to understand at this stage of human development.

Everyone has his or her own definition of what God is, and further examining the nature of God is something well beyond the scope of this book. This work does not attempt to overturn or verify religions, but rather provide a way for us, spiritualists, scientists and others to further investigate and better understand our souls. We are at the early stage of developing our understanding of the soul and how it operates and will undoubtedly learn more and make new discoveries that will further enlighten us.

And if "the brain is the medium of expression of the human mind, similarly the entire physical universe would be the medium of expression of the mind of a natural God. In this context, God is the supreme holistic concept," suggested Burr.[6]

And the better we understand our minds, as the Hindu Bhagavad Gita claims, the better human beings we will be—and so much closer to God. No wonder spiritual experiences are so powerful to all of us.

PART TWO—SPIRITUAL PHENOMENA

17—A New Age of Science

A closer look at science, new age practices, health, medicine and religion provide further insights into the soul and spirituality and how they work—as well as how some things that are considered spiritual are not.

Science aims to describe, measure and explain the world we live in. If two scientists in two different laboratories come up with the same result, you have to take that result seriously. There is only one science and it is the same everywhere and produces the same results. Measurement that can be repeated is the foundation of modern science, and measurements and mathematics that don't appear to work are treated skeptically. Scientists are not swayed by cultural, regional or other subjective influences. That is why any investigation into the soul must include a scientific element.

Yet, when two different religions develop the same philosophy it is often dismissed by scientists as unsubstantiated. For example, the Christian and Buddhist notions of compassion have developed parallel emphasis on love and forgiveness while half a world away from each other. This, and several similar examples, suggest there is indeed something substantive about spirituality.

Today the schism between science and religion is being bridged. There are an increasing number of articles and books on religion and spirituality written by scientists, as well as some on science written by theologians. There is a prestigious prize provided by a financial institution for the best work of the year that bridges religion and science[1] and The Vatican has hosted conventions on spirituality that has included eminent scientists as well as theologians.[2]

In fact, The Vatican has even recommended its clergy consider the findings of science. For example, in 1979 on the 100th year anniversary of the birth of Albert Einstein, Pope John Paul II told a meeting of the Pontifical Academy, "I would like to confirm again the Council's declaration on the autonomy of science in its function of searching for the truth inscribed during the creation by the finger of God. Filled with admiration for the genius of the great scientist, in whom is revealed the imprint of the creative spirit, without intervening in any way with a judgment on the doctrines concerning the great systems of the universe, which is not in her power to make, the Church nevertheless recommends these doctrines for consideration by theologians in order to discover the harmony that exists between scientific truth and revealed truth."[3]

Spirituality and science are two sides of the same coin. Or as Einstein said, "science without religion is lame, religion without science is blind."[4] Einstein also suggested, "Everyone who is seriously engaged in the pursuit of science becomes convinced that the laws of nature manifest the existence of a spirit vastly superior to that of men, and on in the face of which we with our modest powers, must feel humble". He described this as "we are in the position of a little child entering a huge library filled with books in many languages. The child knows someone must have written those books. It does not know how. It does not understand the languages in which they are written. The child dimly suspects a mysterious order in the arrangement of the books but doesn't know what it is. That, it seems to me, is the attitude of even the most intelligent human being toward God. We see the universe marvelously arranged and obeying certain laws but only dimly understand these laws."[5] Similarly, Candace Pert suggests science, at its core, is a spiritual endeavor.[6]

Using science and spiritualy together to understand the world is like using two eyes rather than one, says John Polkinghorne, a physicist and priest.[7] He says the discoveries of science strengthened his convicton that the design of the universe pointed to a divine intelligence, even though science could not explain why it was there in the first place. He adds, "I think that young earth creationism is shutting its eyes to extremely well-motivated conclusions. I'm sad about that because religion is about the search for truth. I think that religious people shouldn't fear truth, from whatever source it may come."[8]

So while in centuries past, science may have undermined religion and spirituality, it is now providing a foundation for them.

This new approach is not only providing advances for both, it also suggests a harbinger of a new age of science. "I think we are on the threshold of finding God—or at least a higher glory," suggests physicist Leon Lederman. "Science is looking in the right direction."[9]

With these advancements, we are also entering a renaissance of the study of the soul and things spiritual—and finding new explanations for things that were previously simply attributed to "faith" or mysticism. For example, our new knowledge of the way electromagnetism is involved in our minds and souls scientifically explains some aspects of mystical new age practices.

18—Making Science of New Age

Native people around the world, who have relied on energy-related healing from their shamans for thousands of years, say that scientists are just now catching up to them in providing scientific interpretations of what was intuitively known.

However, there is still a long way to go and many issues to research. For instance, we regularly hear about minor miracles. Someone is told they have cancer, then a few months later there is no sign of the disease. How were they cured while someone else was not? Similarly, there are reports that someone sensed an event before it happened or that a psychic revealed accurate information about a departed relative that they could not have possibly known. There are too many of these incidents to be mere chance. Is there a logical explanation, as is the case with magic tricks, or a scientific basis as is the case with spoon bending?[1]

There are answers for many things considered spiritual. Let's start by looking at some of the older "new age" practices to see what elucidation they can provide as to our souls and spirituality.

Aruyveda—is an ancient Asian medicinal tradition, which has been practiced in India and Sri Lanka for more than 3,000 years. Through intense meditation, some 52 Rishi prophets of ancient India gained the knowledge or "Veda" of how the world and everything in it works. Their revelations about health and sickness were organized into the system of Aruyveda—or knowledge of health and illness.

Long before western medicine, Aruyveda reportedly linked mosquitoes and malaria, and rats and plague, as far back as the 5th century.

Besides providing insights into the body and medicine, it also provides some insights into the soul and spirituality. For instance, a fundamental of Aruyveda is that everything within the universe is composed of energy or "prana"—including ourselves. This appears analogous to the fundamental energy of string theory.

Aruyveda says we inherit our energy from our parents when we are conceived and are born with three energies. These are:

- a driving force energy that is associated with the nervous system and body called Vatha;

- an energy system that is related to digestion and metabolism called Pitta or fire energy; and
- a water-based energy system called Kapha.

While each of us have all three of these energy systems, we may be predisposed towards one or more of them. Our individual constitutions and bodies are the way and shape they are due to this mix.

The balance of these three energies depends on a variety of factors, balanced spiritual and emotional health and the like. And this energy changes further according to circumstances, such as our lifestyle, our diets and the world around us. Some of these changes can be positive and others negative.

Aruyveda also recognizes that energy controls the function of every cell, thought, action and emotion. This is analogous to the electromagnetic forces detailed in Part I.

In terms of harnessing this power and developing our soul, Aruyvedic tradition says we must live in a way that encourages positive energy, or at least balance the negative with positive energy. It also says that the thoughts we think affects the quality of our energy and consequently our health—and lives. In Vedic philosophy, our lives are believed to become more meaningful when we strive to fulfill our potential.

While Aruyveda does not provide scientific detail as to what this energy actually is and how it operates, it appears to describe something similar, in part, to the electromagnetic aspects of our nervous system, souls and spirituality.

Chinese Medicine—many Chinese similarly believe our bodies have an energy that flows throughout them, a life energy called "chi" or "qi" (pronounced "chee").

People receive chi energy from heaven and earth. Heavenly chi instills an element of the divine within us. We are born with a fixed amount of chi, which is inherited from our parents. We can nourish this yuan chi, though we cannot change or add to it, though we may deplete it by negative practices. Another type of chi is gong chi, which is derived from the air we breathe, food we eat and energy around us.

This chi energy can further be divided into two universal energies, yin and yang energy. Yang chi is said to be more prevalent during the light, while yin energy is more prevalent during darkness. A balance is required between these two energies, says Chinese medicine and philosophy.

The philosophy behind Chinese medicine is that people live between heaven and earth and comprises a miniature universe in ourselves. Doctors using Chinese medicine prescribe a method to correct energy imbalances, be they physical or psychological. Once the doctor pinpoints the excesses that cause the imbalance, they can be adjusted and the patient's physical and/or mental health restored. As such, they treat the person, not just the illness; as illness is a manifestation of an imbalance that exists between a person and their universe. Restoring overall equilibrium to a patient's flow of energy is the ultimate guiding principle of Chinese medicine.

Chinese Taoists, in particular, suggest that as the mind moves, chi energy follows.[2]

Again, this several thousand year old approach of Chinese medicine and philosophy is another indication that an energy system underlies ourselves and imparts divinity and spirituality to us.

Acupuncture—Chi is said to flow through the body via a system known as ching-luo, jing-luo or main and collateral channels, which include the nervous, blood and other bodily systems. At certain points it is funneled, concentrated and sometimes restricted or blocked. It is at these points where acupuncture needles are inserted to relieve the energy blockages.

Acupuncture was listed as an effective treatment of 43 types of ailment by the World Health Organization in 1980. A study at the University of California's Irvine Medical Center also proved that acupuncture works in animals. This study is important, as animals are not believed to be subject to the placebo effect. It found the release of opiates in the brains of cats resulted from acupuncture.

Scientists suggest acupuncture produces analgesia, or loss of pain sensation, by increasing the release of opoids.[3] It is thought that opoid peptides are the body's natural painkillers. Opoids are a type of neuropeptide cell and include the amino acid enkephalins, as well as endorphins and dynorphins. They can have pain-relieving effects 200 times stronger than the effects of morphine. Also, they have also been linked to generating feelings of euphoria, sexual drive, improved learning and memory and a range of other body regulation functions.

It is also known that when low frequency electrical stimulation is applied to acupuncture needles that the pituitary gland is also aroused and activated. The pituitary gland is believed to be a major command center of the brain. When high frequency electroacupuncture is applied, the neurotransmitters acetylcholine, serotonin and dopamine are released into the brain. These studies suggest there is a chemical nature to acupuncture.

One scientist found that about 25 percent of the acupuncture points on the human forearm do seem to exist, in that they have specific, reproducible and significant electrical parameters and could be found in all subjects. Robert Becker, in his book *The Body Electric*, concludes that the acupuncture system is actually there.

Becker suggests acupuncture works by the needles initiating direct current electric signals that carry information that an injury has occurred via acupuncture meridians to the brain, where these signal are consciousnessly perceived as pain. "The remainder went to more primitive portions of the brain, where they stimulated similar output DC signals that caused the cells and chemical mechanisms at the site of injury to produce repair. This is a complete closed-loop, negative-feedback control system."[4]

Becker also found acupuncture points and meridians conduct electric current and flow into the central nervous system. Each point is positive compared to the surrounding area and has an electric field surrounding it. "We even found a 15-minute rhythm in the current strength at the points, superimposed on the circadian ('about a day') rhythm we'd found a decade earlier in the overall direct current system. It was obvious by then that at least the major parts of the acupuncture charts had, as the jargon goes, 'an objective basis in reality'." He suggests the structure that carried the current, in a manner that did not interfere with nerve impulses, was the system of perineural cells.

In short, the classical acupuncture system, at least in part, has unique electric characteristics with several acupuncture points known to have lower electrical resistance and involve the continuous current nerve system.[5]

Herbalism—both Aruyveda and Chinese medicine use a range of herbal remedies. Such remedies have not only been around for millennia, but some have also been prescribed as benefiting the soul.

Some plant forms are known to be very therapeutic, such as the mould of penicillin, and form the basis of modern Western medicines.

However, studies into others, such as St. Johns Wort and Echinacea, have demonstrated mixed results regarding their therapeutic effectiveness. Some studies have shown that Echinacea does help stimulate the human immune system, while others show it does not. Which studies are correct? They might both be, as there are three different species of Echinacea that are used medicinally. Herbalists say their effective varies on the species of plant, how fresh it is, where it is grown, how it is harvested and a range of other factors.

In another study, a compound extracted from sage has been shown to help people with Alzheimer's disease, due to the blocking of an enzyme linked to the disease. By blocking the enzyme, researchers at King's College, London, stopped the breakdown of a chemical messenger called acetylcholine. Low levels of acetylcholine lead to gradual loss of short-term memory.

On a more common level, take the common garden carrot. Carrots are the richest common source of beta-carotene, which the body converts into vitamin A, which combines with a protein in the retina, helping improve sight, compared to those who are short in vitamin A.

While such studies suggest herbs work on a chemical level, there may also be an energy element. Some people suggest there are 'positive' as well as 'negative' herbs as well as foods. Positive ones such as fruit and vegetables are thought to stimulate the production of dopamine, or feel good chemicals, in the brain, while fatty and sugar-laden foods can have the opposite affect. Yogic knowledge says foods trigger certain reactions, with some foods stimulating, such as acid-forming foods like red meat; while others are calming and tend to include alkaline foods, such as fruits and vegetables. However, brain specialists caution there are a range of reasons why chemicals are produced in our bodies and that you can't necessarily cure someone with depression by feeding them more fruit, though it might help.

While on an outback adventure in the middle of the Australian dessert, I met a man who claimed that he could tell what was a beneficial or a harmful plant. This man, Brian, who graded a dirt road hundreds of miles from any town, would hold part of a plant in his left hand and lift something heavy with his other hand. If he felt stronger, he considered the plant beneficial, while if he was not able to lift the load as well as normal, he said the plant contained negative energy (and should certainly not be eaten).

Valerie Hunt reports something similar. "A simple test for 'good' or 'bad' field interaction is demonstrated with applied kinesiological manual testing of the muscle strength of the arm when a single food, herb or medication is placed in one's field or held in one's hand. If the muscles test strong to downward pressures on the arm, the field interaction is positive and that substance is not harmful to the person. If the muscles become weakened and the arm cannot resist pressure, the field interaction is poor and the substance should be avoided," she says.[6]

This would suggest a kind of wave coherence and/or stochastic resonance; that the wave energy of a herb or food contributes to heighten the sensitivity of some other aspect of our physiology.

Aromatherapy may work on a similar level, invoking additional energies through the sense of smell.

In a related area, Bach flower remedies are said to work because they contain the 'life force' of the flowers used to make them and are believed to work on a vibrational energy level to heal mentally and emotionally. The developer of this practice, Edward Bach believed illness was a manifestation of a deeper disharmony within the energy of a person. However, studies in this area currently inconclusive.

Special waters—there are several naturally occurring baths and springs in Europe and many other parts of the world that are proclaimed to provide miraculous healing powers to those who bathe or drink from them. Is it possible for water to create miracles or are these just other examples of the placebo effect?

Water is the most important and abundant inorganic compound in all living things on earth. It comprises most of the human body and is where many chemical reactions occur. There are 10,000 water molecules in the human body for every molecule of protein.

One of the most important physical properties of water is its polarity (or a partial negative charge near the one oxygen atom and two partial positive charges near two hydrogen atoms). This makes water an excellent solvent for other ions and provides it with a greater electric charge than many other molecules. This not only enables water to dissolve many molecules that are important to life but also efficiently convey electric currents.[7]

Interestingly, low frequency electromagnetic waves travel through water better than higher frequencies (which do not penetrate as well). This obviously makes water an important element in terms of energy flows within our bodies. This could be one reason why a hot shower or bath feels so good.

Besides these fundamental properties, most of the water in these miraculous baths and springs contains increased amounts of minerals; inorganic chemicals which are also necessary for many biochemical processes in our bodies. Consuming them could restore imbalances within the bodies of those who drink them, especially if they are deficient in one mineral or another.

Some of these waters also include increased amounts of magnetic minerals, which we know can have some effects on our electric flows. This also could explain how these baths make some people feel better, though the exact mechanism for this is not yet understood.

Another benefit from such waters if they are consumed could be that they contain alkalizing minerals, which we know counteract acidity in body pH. Too much acid in the body has been blamed for a range of diseases, including osteoarthritis, osteoprosis and several other ailments.

Acid is created by carbon dioxide, a gas that is produced in high concentrations in our bodies as a result of metabolizing food. Carbon dioxide, a partially charged gas, reacts with water to form toxic and destructive carbonic acid. This acid affects the shape and activity of enzymes and other proteins in our cells as well as change the pH balance of body fluids, which impacts the ion potential gradient of mitochondria, in turn reducing the amount of energy produced in our cells—in fact, the amount of energy produced in our bodies, leaving us feeling fatigued. Then there is the impact of carbon dioxide itself on brain cells, where it is known to be associated with Alzheimers disease.

Another interesting aspect of carbon dioxide in the body is that it is reduced by reducing calorie intake, which could explain why decreasing calories has also been equated to increasing life expectancy.

We produce an estimated pound or two of this gas a day, which we have to get rid of. Breathing is obviously a very important way to do this, as is drinking water. This can help reduce the amount of acid and restore the body's pH balance.

Reducing acidity in the body through the correct balance of minerals in water can improve health and extend life, according to research by Australian biochemical pathologist Russell Beckett. He noticed that sheep and cows in one part of Australia lived 30–40% longer and remained healthier than their counterparts elsewhere and found the only difference was the composition of the water they drank. A major difference in the water was specific concentrations of acid-reducing magnesium bicarbonate.

Magnesium bicarbonate not only reduces the amount of acid in the body but also increases the energy output of our cells. Beckett says, "Magnesium carbonate increases considerably the energy production in body cells. This energy increase occurs in several ways. First, magnesium bicarbonate protects the natural organic and inorganic phosphate buffers in the cytoplasm of cells. This is important, particularly in muscle cells and brain cells (neurons). Second, magnesium bicarbonate neutralizes the acid produced as a result of metabolic processes and ATP (adenosine triphosphate) hydrolysis. This allows more ATP to be produced. When more ATP can be hydrolysed and more ATP can be produced, body cells have sufficient energy for optimum function. When body cells have optimum function, the energy levels and the physical performance of the body are enhanced," Beckett says.[8] Magnesium bicarbonate also helps maintain the optimum electrochemical gradient across cell membranes, he says. In this process, carbonic acid is broken down by the body into positively charged hydrogen, which we have seen earlier is required for ATP and, negatively charged HCO_3^-bicarbonate ions, which are the second most prevalent extracellular anions (negatively charged ions) after potassium in the body.

Interestingly, Beckett also found that the growth of some cancers is related to high acidity in the body. "Most cancers derive from epithelial tissues where carbon dioxide concentrations are high and acidic conditions are common. Cancer, or anaplastic cell growth, can be interpreted as a form of cell survival in response to acidic and other adverse environmental conditions. In addition, the acids inside lysosomes are considered to be necessary for the metastatic invasion of some cancers. The lysosomes cause tissue destruction, which allows the cancer cells to invade. Indeed, one scientific theory of ageing and fatigue states that lysosomes may contribute to constant body destruction and this may be responsible for the visible appearance of ageing and for physical ageing and fatigue per se. This explains also why old cells and tissues appear more vulnerable to pathology and disease than young cells and tissues. That is, in old cells and tissues, there has been a greater exposure to the acids and destructive enzymes of lysosomes. The ability to maintain cell and tissue function is impaired," he theorizes.[8]

So far, only limited medical trials have been conducted with magnesium bicarbonate in people with a range of degenerative, inflammatory and viral diseases, Beckett says. "The results of all trials were unequivocal. People who consumed at least one and a half liters of water that contained appropriate mineral complexes of magnesium bicarbonate experienced a range of health benefits. In general, people who consume magnesium bicarbonate have remissions in the clinical signs of those diseases that require carbon dioxide concentrations as part of the disease pathogenesis." He has since developed and sells a special form of mineral water, which he calls Unique Water and secured both US and Australian patents for this water. These are believed to be the first patents granted for treating aging and increasing human life span.

This work by Beckett, and others have found the mineral content of these special waters is the only difference between them and ordinary tap water. "There is a library of scientific knowledge that communities which drink hard (mineral rich) water tend to live longer than communities which drink soft (mineral free) water," Beckett notes.[9]

This indicates there is nothing spiritual (other than a belief or placebo element) in this respect, but rather an important chemical process. It shows how important chemistry can be to our biology.

Homeopathy—this practice dates back to ancient China, where doctors knew to take scabs from the sores of smallpox victims and rub them into cuts of other people to protect them from the disease.

In the west, the practice's 18th century founder Samuel Hanheman believed that by diluting or "successing" certain plant, animal and mineral substances in alcohol and water he was releasing the energy from within them. He used these solutions to treat symptoms, in particular recreating a symptom to counter the symptoms created by a disease. This is described as creating an energy opposite to another so that the two cancel out.

Interestingly, Hanheman believed that we all have our own energy, our own "vital force," which stimulates our body, mind and emotions. It can be disrupted by a range of factors, including pollution, stress, lack of exercise, poor diet, and hereditary factors and so on. Hanheman designed his homeopathic remedies to stimulate that force to enable to body to be healed.

But how can something that has been diluted (sometimes 100s of times) to the point where it does not contain any of the original substance still work? And why does homeopathy work for some people and not others? Some potential answers to these vexatious questions also provide further insight into the soul. This might be similar to DNA and holograms which, from just a small piece of information, can recreate a larger whole.

Italian physicist Emilio Del Giudice claims substances have radiation fields of charged molecules which interact with water molecules to create polarized water, or water that he says still has the "energy signature" of the original substance. He suggests this signal is transferred to the physical body, via homeopathy, where the information creates change, or remedy.

French scientist Rolland Conte and his team suggest homeopathy works through "remanent waves" and what they label "white holes" (as in contrast to black holes in the universe). They say water has some form of memory due to waves created by the protons of the original substance before it is diluted out of the solution. So while the substance has disappeared, the wave has not and remains as a remanent. They claim their experiments involving protons of hydrogen and isotypes of oxygen indicate a reorganization of the constituents of the dilued water and lead to the appearance of a new quantum state which has not been totally explained as yet, but which is consistent with wave theory of quauntum physics.

Another French scientist, Jean Benveniste, notes "that molecules vibrate, we have known for decades. Every atom of every molecule and every intermolecular bond, the bridge that links the atoms, emits a group of specific frequencies. Specific frequencies of simple or complex molecules are detected at distances of billions of light years, thanks to radio telescopes." Vibrations "send instruction to the next molecule down the line in the cases of events which govern biological functions, and probably, to a large extent chemical ones as well. We could transfer specific molecular signals using an amplifier and electromagnetic coils. In the course of several thousand experiments, we have led receptors (specific or complex molecules) to 'believe' that they are in the presence of their favorite molecules by playing their recorded frequencies of those molecules."[10] He suggests water can record the electromagnetic signals of various molecules, even when those molecules are no longer in the water. In other experiments, he has used low frequency (less than 20kHz) electromagnetic waves to activate specific cell functions.

Other scientists dispute his findings: some have been able to repeat them, while others have not—leading to controversy surrounding this work.

Interestingly, Benveniste adds the effects of homeopathic solutions are erased by a magnetic field. The phases of the moon and regional geomagnetism are also known to impact the quality and effectiveness of homeopathic products and results.

Conte's team also found that gravity not only affects the quality of homeopathic solutions, but also growth. They found that growth, in the case of vegetables, is strongest when gravity is weakest. This gives some credence to the old farmers' adage that for rising seeds, seeds which grow towards the sky, plant from full moon to waning moon. Conte and his team suggest this is due to the fact that space-time is less deformed when gravity is less.[11]

These findings indicate there is a strong electromagnetic component to homeopathy.

In contrast, others note that homeopathy is symptom-orientated and focuses on body's reaction, rather than on the cause of the problem, and that homeopathy's benefits are primarily due to the placebo response. Psychotherapist Dylan Evans says the secret ingredient in homeopathy is belief.

The jury is still out on how homeopathy works. Further research, taking into account electromagnetic affects, is required.

Magnetism—Electromagnetic fields contain both an electric field and an associated magnetic field, with the latter generated by the electric flow. If the electric current and voltage changes, this adds a wave or directional aspect to the field. This wave carries information and energy that is able to influence other parts of the field.

Magnets have been claimed to cure all sorts of ailments. The north pole of a magnet is said to be equivalent in energy to yin energy of Chinese medicine or Aruyveda's shakti energy and is claimed to produce inhibiting effects such as quietening, soothing and cooling. The south pole of a magnetic is said to correspond to yang or shiva energy, which is said to produce faster and warming effects. Others claim a negative magnetic field is claimed to increase cellular oxygen, pull fluids and gases, reduce fluid retention, encourage deep sleep, fight infection, support biological healing, reduce inflammation, reduce acid, relieve pain and promote mental acuity.

Science has found that electrons from the north pole of a magnet move counter-clockwise, while those emanating from the south pole spin clockwise.

Many scientific studies have shown no benefit at all from magnets.[12] Magnets do not bind iron in the blood and they do not create warmth.

This does not mean that magnetism does not have an affect on cells. Studies have shown that magnets can affect the way calcium binds to certain molecules, speed up the release of some enzymes, increase the solubility of some chemicals in water, as well as ion concentrations and pH.[13]

To try to resolve the controversy, a substantive study was undertaken by researchers at the University of Virginia into the effects of magnetic therapy for pain relief. It found no significant statistical difference in treatments using magnets or not. However, it did find a difference in the level of pain, with people who slept on pads containing active magnets reporting the greatest reduction in pain and tenderness. Another study also found that permanent magnets reduced pain in post-polio patients.[14]

Adding magnetism to an electric current can influence it, but only if enough magnetic energy is applied. If you want to divert an electric current, you need very powerful magnets. Ampere described magnetism as "electricity thrown into curves" and this is just what powerful magnets appear to do. This diversion of electric current by magnetism would only be useful if there was a medicinal or health need to reduce energy flow or electric signaling, such as to and from a painful area. In reducing the pain signal, magnetism could also disrupt signaling required for healing.

Most human tissue is weakly magnetic. In fact, most tissue exhibits a slight tendency to become magnetized in the opposite direction to an applied magnetic field—what is called diamagnetism.[15] Diamagnetism distorts the orbits of electrons in atoms, generating a tiny electric current, which generates a magnetic field in the opposite direction from the main magnet. Diamagnetism, using very strong magnets, was how a frog was levitated in mid-air by Andre Geim, a physicist at the University of Nijmegen in The Netherlands.

There is another way that magnetism might affect people. The pineal gland, often described at the "third eye" by mystics, produces melatonin and serotonin, two neurohormones that control the body's biocycles among other things. Researcher have found that by slightly increasing magnetism around the pineal gland increases production of these neurohormones, while decreasing the magnetism can also decrease their production.[16] Subjects who are insulated from magnetic fields develop longer and more irregular biorhythms. Researchers have also found another magnetically sensitive chemical, magnetite crystals, elsewhere in the human brain.[17] It is believed that these magnetically sensitive molecules enable the brain to detect small magnetic fields.

Otherwise, the impact of magnetism alone appears to be limited in the human body's somewhat insulated electromagnetic field.

Feng Shui—is based on the arrangement of different elements and their interactions with energy. Original Feng Shui texts are formulated on the flow of energy, in particular ions. It uses symbols and metaphors such as dragons and tigers to describe negative ions flowing from mountains through the air towards water, which is an ion-sink, along with other natural forces such as metal and wood which can stabilize such flows. Today's Feng Shui has become a much more metaphorical rather than science-based practice.

Touch—if we can use the mind to assist physical healing, maybe we can use physical attention to heal our minds—and maybe spirit.

There are numerous reports of people being comforted by the touch, hand-holding or hug of another person. When a child cries, a cuddle or a hug is every parents' remedy around the globe.

Even when unconscious, as when anesthesia, a patient's heartbeat and blood pressure can be calmed and lowered by a comforting touch.

Ancient societies record how massage was used to treat diabetes, while various healings are described in the *Bible,* and the "laying on of hands" used by Jesus is still used in some churches. Was Jesus an early practitioner of therapeutic touch?

However, since Freud there has been a tradition of no-touching when dealing with issues of the mind and spirit. This has been adopted and extended by some religions,

In contrast, science is now showing that touch can work certain wonders. For example, studies have shown that touch can be useful in lowering blood glucose levels and encouraging relaxation.

A Ohio University study of heart disease in the 1970s found that one group of the test rabbits fed artery hardening food displayed 60% fewer symptoms of blocked arteries and the like, but for no apparent reason. That was, until it was discovered the handler in charge of feeding these lucky rabbits would fondle and pet each one for a few minutes before feeding. This human touch and affection was credited with reducing the affect of the toxic diet provided to all rabbits in this test. Repeated experiments, where one group of rabbits was similarly petted and another group was not produced similar results.[18]

Is this why people often feel better after a hug or massage? How can it work?

The electric field given off by the healer induces electromagnetic resonance of some body component, suggests Becker, and the healer senses the returned signal.[19]

He is supported by researcher from the Institute of HeartMath. "When two individuals touch or are in proximity, one's electrocardiogram (ECG) signal is registered in the other person's electroencephalogram (EEG)," notes the institute's Rollin McCraty. "While the transmission of the signal is strongest when people are in contact, the effect is still detectable when subjects are in proximity without contact. Our results suggest that the signal transferred is electromagnetic in origin and that some component of it is radiated. We have also found the degree of coherence in the individuals' cardiac rhythms to be an important factor in determining if biological synchronization occurs between the two subjects, especially when subjects are separated by larger distances. This study is one of the first successful attempts to directly measure an energy exchange between people, and provides a testable theory to explain the observed effects of many healing modalities."[20]

Pert adds that receptor molecules respond to energy and chemical signals by vibrating. "Receptors and ligands attract at a distance as they resonate at the same frequency. Perhaps we can now begin to imagine how physcial 'adjustments' of spinal joints that house petidegic nerve bundles, therapies that emphasize emotional expression and feeling within the body, and hands on healings where practitioners claim to be able to feel the energetic differences and emit appropriate corrective energies share common energetic mechanisms."[21]

Research by James Oschman found low frequency pulsed electromagnetic fields emanate from the hands of some touch therapists.[22] Other research claims Qigong masters emit an infrared energy from their hands of between 3–5 micrometers wavelength.[23]

These studies scientifically provide a framework as to how touch can impact others, through wave coherence.

Of the various touch therapies, Jin Shin Jyutsu accords the most with scientific principles. A little like acupuncture without the needles, it uses low frequency energy from the practitioner's hands (rather than triggering opiates with acupuncture needles) to balance energy in the body. Using touch to connect certain points on the body, the touch of Jin Shin Jyutsu "jump-starts" a person's energy flow and removes blockages to restore flow. The right hand is used to "send" energy, while the left is used to "receive" it.

Indeed, if the energy field of one person is similar to that of another being treated (in terms of frequency, patterns and so on), and the principals of parallel electronic circuits apply, such touch therapies could provide more energy to the subject and even be beneficial in some circumstances. For example, light bulbs lit via a 'series' electric circuit do not burn as brightly as those powered by a parallel circuit.

Interestingly, the woman who brought Jin Shin Jyutsu from Japan to the West, Mary Burmeister, interestingly described "matter is the lowest level of spirit, and spirit is the highest degree of matter".

As we all can testify, the touch of a person can have a tremendous influence our own bodies and mind.

Meditation—Meditation is known to be able to produce slower breathing, slower heartbeats and slower brain waves.

People who meditate tend to absorb less oxygen from the air when they breathe, compared to when they are not meditating. This suggests their whole body relaxes and uses less energy, according to early research by Robert Keith Wallace in the 1960s. He found that practiced meditators reach a level of lower oxygen consumption than when they are asleep. In sleep it can take several hours to reach such a period of low oxygen consumption, whereas it can be obtained much quicker during meditation.

Wallace and Herbert Benson of Harvard University also found that practitioners of transcendental meditation reduced stress in their lives, recover faster from illness, slow aging, and improve creativity and productivity. Wallace found meditators tend to have bodies that are biologically younger than their calendar age, based upon tests on sight, hearing and blood pressure.

This has since been supported by another study, which found people who meditate tend to be healthier than the average American is. The sizeable study by health insurance firm Blue Cross-Blue Shield found that 2000 people who meditated were hospitalized 87% less for heart disease than those who did not.

In another study, elderly people taught meditation were more likely than their peers to still be living three years later. Meditation proved more effective than two other techniques for longevity, lowering blood pressure and improving on several measures of mental functioning, researchers at Harvard University found. In their study, 73 volunteers, with an average age of 81, from eight homes for the elderly were assigned randomly to learn transcendental meditation or no training at all. Three years later, the 20 taught Transcendental Meditation were still alive. Survival rates in the other three groups were 88 per cent, 65 per cent and 77 per cent, respectively.[24]

However, another study found that meditation produced no changes in the level of stress-related chemicals in the blood compared to a control group that just sat quietly.[25]

An analysis of brain scans of Tibetan Buddhist monks found that when they meditate their brain waves diminish in an area that helps people to orient themselves in three-dimensional space. This could explain the feeling of many people who meditate that they feel at one with the universe. Andrew Newberg and Jeremy Iversen said their study showed increased blood flow in the bilateral frontal cortices, cingulated gyrus, and thalami of the brain. At the same time, a decrease in blood flow was recorded in the superior parietal lobes, with the left more affected than the right.[26]

Controlling and reducing brain waves could also generate coherence between brain, heart and other body waves to provide the sensation of relaxation and pleasantness. Recall the work of the Institute of HeartMath in this respect.

Oschman explains how imaginging in active meditation may work. "First, mental practice of movements sets up the anticipatory fields described above without causing any muscles to move. Kasai et al (1997) refer to this as 'subthreshold muscle activity'. We would expect imagery to produce the readiness potential and, possibly, the pre-motion potential, without the motor potential that triggers the movement. Mentally reheasing an action sends information throughout the body, via the perineural and other conductive systems, to all of the relevant cells. This then leads to a 'preconditioning' of biochemical pathways, energy reserves, and pattersn of information flow. Cells everywhere are then poised to worth together at the instant of demand."[27] This practice is used by many elite sportspeople, so why not for the rest of us if given a little practice. It is also most likely that a similar process is inolved when it comes to "positive thinking", as mentioned earlier.

Yoga—the *Yoga Sutra* written centuries ago by Patanjalia notes the goal of yoga is "the resolution of the agitations of consciousness". It achieves this through the controlled "expulsion and retention of breath".[28]

The origin of Yoga is also intertwined with Auryveda, with different body poses designed to influence and balance the different types of body energy identified by Auryveda.

More recently, various studies have found that relaxation techniques of sitting, visualization and yoga, yoga provides the greatest increase in alertness, mental and physical energy.[29] Research has identified that during yoga, peptides and other biochemical messenger molecules are released from the brain.

This might explain reports that yoga can reduce pain and influence behavior and attitude. As one Indian researcher notes: "Yoga primarily changes your consciousness, which includes your way of looking at things. In the process, many aspects of your physical functioning also change, including your brain chemistry."[30]

Chant and Music—rhythmic stimulation, be it controlled breathing, chanting, drumming, praying or swaying, have also been shown to alter the electrical patterns of the brain.

Any repetitious movement evokes a change in a person's physiology that counters stress-induced adrenaline and helps the body relax, says U.S. cardiologist Herbert Benson. He adds it does not matter what that repetition is, be it repeating a sound, a word or phrase or prayer over and over or a rhythmic exercise. "There are two steps to breaking the pattern of stress: repetition of a word or activity, and, when other thoughts come to mind, passively disregarding them and coming back to repetition. It is your choice whether it be religious or secular," but it must be based in your personal belief system to bring about relaxation, he cautions.

Benson adds, "I then studied the world's secular and religious literatures and discovered that these two steps were present in virtually every human culture. The earlier account I found is Hindu; it speaks of individuals focusing on their breathing and repeating a phrase of scripture on each out breath while disregarding everyday thoughts. The same concept was employed in one practice of early Judaism. In Christianity, prayers codified in the 14th century taught people to focus on their breathing while repeating the prayer "Lord Jesus Christ have mercy on me". We found the same pattern in Islam, Shinto, Taoism and Confucianism. Only the words differ."

In fact, chants are well established in the religious practices of many countries. There is the chant "om" in Tibet, while in China there is "namo ammitofo" or "I take refuge in the Buddha". For Christians it is often "Our father who art in heaven" or the Hail Mary for Catholics, while Muslims often cite "Insha'allah" and Jews "Sh'ma Yisroel".

The Gyuto monks of Tibet have a special way of chanting that appears to build on each previous note, amplifying the vibration of the previous note/sound—creating a resonance. Some people claim this resonance is a spiritual technology that allows these chanting monks to "tune into" a divine wavelength.

It is as yet uncertain if chanting creates a coherent resonance of brain, heart and other waves, or weather it also involves stochastic resonance at some stage to allow meditators to become sensitive to previously undetected stimuli or unconscious thoughts (brain waves that have not become strong enough to become conscious thoughts).

Besides altering the brain patterns of the brain to some extent, chanting also appears to have a chemical element. Yogic knowledge, for example, claims there are 84 energy points on the roof of your mouth that connect to the pituitary gland. When you chant a mantra or sound, these points are stimulated in a certain permutation and combination that sends a pulse to the pituitary gland, which in turn releases various chemicals and ultimately, feelings.

Researchers note that chants can activate the pituary gland to secrete various chemicals into the brain such as endorphins, acetylcholine, serotonin, dopamine and norepinephrine, which trigger other chemical reactions and ultimately produce pain relief and relaxation throughout the body.

Chants also appear to be able to work at another level, that of belief and perception as in the case of people repeating positive affirmations to try and alter their perceptions. However, repeating positive sayings in an attempt to channel positive attitudes into oneself has widely varying results, especially as a treatment for disease. One reason for this is that not everyone who repeats such affirmations truly believes them; they may still harbor negative feelings, and accordingly energies, underneath them that counter them.

Music, like chants, can also set up harmonic vibrations that can be beneficial. Music therapy has been demonstrated to be beneficial in terms of stress reduction, pain relief, improve some aspects of immunity, cardiovascular and endocrine activity.[31]

Music is known to be able to alter emotions, with researchers in Japan showing that music can affect the limbic system. Whether this is direct through vibration or via perceptions is as yet uncertain.

One possible explanation is that sound waves can cohere and smooth some of our own electric waves (in our hearts and/or heads), while the words and tunes appeal to the mental part of ourselves, the combination of the two eliciting a response from the soul.

Interestingly, by listening to the same musical notes, but presented in differing rhythms, researchers at the Massachusetts Institute of Technology (MIT) found subjects reported the sounds seemed fresher, even though they were the same notes. An interesting side effect was that when the subjects went home they reported colors appeared brighter and food was tastier. It was as if all their senses had been heightened by simply changing the rhythms (speed pattern) of the notes.

Hearing is an interesting sense, with many as yet unanswered questions. For example, why is the frequency range of what we can hear 10-fold the frequency range of what we can see? Why can we close our eyes, but not our ears? And why is hearing generally the last of our senses to leave us at death?

Astrology and numerology—the fact that astrology and numerology have been around for centuries, suggests there is some basis to them, even though there is little scientific evidence to support their practice.

In one interesting experiment, British astrologer John Addey recorded the birth times of more than 7,000 doctors and clergymen and sorted them according to the location of the sun in space at the moment of their birth. Addey found that many of the doctors' birthdates were grouped every one-fifth, 25th and 125th of the way along the circle of the zodiac signs (beginning at vernal point of 0° Aries). According to numerology, five is the natural number of the healer.

In contrast, many other scientific studies have found no such accuracy of prediction.

Some sort of seasonal explanation could be applicable, as we know that various electromagnetic forces change with the seasons and can have an affect on people. For example, nerve impulses travel at different speeds during different seasons, at least in the nerve cells of giant squids. And this change is not due to changes in temperature.[32]

The major change that occurs during the seasons is the earth's positioning relevant to the sun, in particular the gravitational relationship between the two. Astrologists have always claimed the alignment of planets have an impact on our lives. Maybe astrology describes differing effects and influences of gravity upon people. Obviously, further objective research is required in this respect.

Auras and Spirits—interestingly, many religions describe their central divinity figure as being surrounded by a bright aura of light, a light associated with spirituality.

Hunt correlates auras of light with the direct current electric fields in our bodies. In determining this, she used innovative research combining not only scientific EGM recordings, but also people who were reported to be able to see auras, to reveal this connection.[33]

Becker notes that the electrical currents flowing in living organisms should produce magnetic fields, fields that extend outside the body. However, this should be extremely weak and generally unnoticeable.

Scientists at the University of California at Los Angeles in the 1980s developed a technique to photograph energy believed to be emanating from humans and other objects, a technique called Kirlian photography.[34] However, the aura captured is described by Kirlian Photography Research (the firm that sells the equipment) as very different to what is often referred to as a person's aura. "What is called an aura by those who have first hand experience of seeing them is not the same as what is detected with Kirlian photography. There may be a relationship, but there is not an identity. What is captured on Kirlian photos is dependent on the high voltage discharge of the device. Ten to forty thousand volts are applied to a piece of film. This electricity interacts with whatever touches the film, and then produces electromagnetic waves from the low infrared to above the x-ray spectrum. This can be viewed directly, or captured by film. It is no surprise that as the high energy from the Kirlian device interacts with the subtle biological system that touches the film, traits of energetic patterns are exposed by the process. The Kirlian aura is very much affected by awareness and health, however auras seen by people are not dependent on the application of high voltage," the firm says on its website.[35] This photography is actually the result of a high voltage charge passing via a metal plate through photographic film to the object on top. Where the charge jumps from the metal plate to the object it collides with nitrogen atoms in the air and excites them and they emit light. This is the luminescence captured in Kirlian images that give the corona-like images.

While there are many spurious claims for other auras, energy spirits and ghosts, science has yet to corroborate these claims. Recall how Persinger says electromagnetic anomalies account for all sensed images. There "is not a single case of haunt phenomena whose major characteristics cannot be acommodated by understanding the natural forces generated by the earth, the areas of the human brain stimulated by these energies, and the interpretation of these forces by normal psychological processes," he says.

However, in another experiment, Arizona scientists explored how accurate "mediums' were in conversing with peoples' dead relatives. The experimenters selected five renown mediums and found they not only did express knowledge of person's dead relatives, but had an 83% accuracy in doing so, compared to just 20% for a control group of "non-mediums".[36]

Carl Jung suggested, "The souls of the dead "know" only what they knew at the moment of death, and nothing beyond that. Hence, their endeavor to penetrate into life in order to share in the knowledge of men. I frequently have a feeling that they are standing directly behind us, waiting to hear what answers we will give to them, and what answer to destiny. It seems to me as if they were dependent on the living for receiving answers to their questions."[37] Jung told a friend that he had come to understand his archetypes were spiritual beings, but thought it best not to reveal it publicly.

Future research will no doubt provide more insight into these and other spiritual phenomena. However, as can be seen from the above, while some do appear to invoke and involve the spiritual, in particular electromagnetic waves, others are based on biochemical processes and reactions and have no spiritual, no mystical, element. Remember when you hear extraordinary claims, you need extraordinary proof.[37]

19—Self Health

If the precepts of Aruyveda and Chinese medicine alone are correct, spirituality is not only important in providing meaning for us, it is also important for our health.

This was echoed by the Greek philosopher Plato when he wrote "if the head and body are to be well, you need to begin by curing the soul".[1]

Now we have a framework outlining the electromagnetic holographic field of the soul and spirituality that can help in terms of these other areas. Can this framework explain miracles in the body? For instance, can it explain reports of someone who is told they have cancer, then a few months later show no sign of the disease? How were they cured while someone else was not? Everyone's body can heal a cut, yet it seems that only a few people can cure a cancer. How can such apparently spontaneous remissions of diseases, or miracles, be accounted for?

Norman Cousin, in his *Anatomy of an Illness* implies that the power to heal oneself is based in how a person interprets illness, that each person's illness is in part constituted by how they interpret their suffering. The meanings they attach to their illness may be as important as the underlying physiological aspects themselves, he says.[2] Similarly, Hippocrates, the founder of medicine in ancient Greece, suggested it is more important to know what sort of person has a disease than to know what sort of disease a person has.

Medical doctor and author, Deepak Chopra, says research on spontaneous cures of cancer has shown that just before the cure appears almost every patient experiences a dramatic shift in awareness. The patient believes they will be healed, and feels the force responsible for that healing is inside them, Chopra says. At that moment, these patients apparently rise to a new level of consciousness that prohibits the existence of cancer. This leap in consciousness seems to be the key, he says.[3]

Such reports suggest perceptions and belief play an important role in not only our spirituality but also our overall health.

Another medical doctor, Robert Becker suggests a more scientific explanation. "Whenever a patient develops a profound belief in the efficacy of any treatment, he or she is in a state resembling self-hypnosis and has access to the DC electrical control systems" of the body.[4] In this way, the immune system is activated, the direct current control system is accessed and tissue healing is accomplished, he says.

Hunt adds that a healthy body, mind and soul incorporates a flowing, interactive electrodynamic field. Anything that interferes with this flow has detrimental effects, she says.[5]

Interestingly, scientists have also discovered that cancer cells conduct electricity better than other cells, with cancers conducting electric currents 7 to 8 times more than normal tissue.[6] However, the energy of cancers are frequency specific with some growing at one electric frequency and others at another. Human cancer cells exposed to 60Hz fields grew 24 times as fast as unexposed cells, according to scientists at the Cancer Therapy and Research Center in Texas. Other studies have also shown that weak microwave fields can enhance the action of phorbol esters, a chemical that is known to grow cancers. Yet, in the absence of the phorbol ester, the microwave had no effect.

This can be used to some advantage. Meryl Rose, for example, found that by passing a specific electrical current through cancers in salamanders that he could get some of the cancer cells to revert back to normal cells and become part of a healing and benign blastema. Rose's work has since been replicated, but unfortunately it has not been pursued among the wider medical community or commercially.

Some Chinese doctors claim improved rates of cancer treatment by injecting platinum into cancers and then passing currents through the cancer cells, destroying them.[7] Similarly, Becker found he was able to cure "incurable" wounds (not cancers) by using silver ions combined with an electric current. He notes: "It is important to realize that this is not simply an electrical effect, but the result of the combined action of the electrical voltage and the electrically generated silver ions. It is an electrochemical treatment. While we do not have firm evidence at this time, what probably happens is that the silver ion I shaped so as to connect with some receptor on the surface cancer-cell membrane. After that connection is made, an electrical-charge transfer sends a signal to the nucleus of the cancer cell that activates the primitive-type genes, and the cell dedifferentiates. In that state, it awaits instructions as to what it is to become. The process is exactly the same as that in Rose's experiments, except that in this case the dedifferentiation is caused by the unexpected action of the positive silver ions."[8] This technique obviously requires more study.

Different types and patterns of electric current being applied; pulsing, direct current, low or high electric field current, all produce different results.

It is the same with our bodies, minds and souls; with different electromagnetic currents producing different results. Recall how the work of Persinger generated spiritual experiences in people, while other experimenters generated more mechanical actions, such as the jerking of a thumb or other part of the body.

The way we think and feel can alter the electric waves within us and how we respond, spiritually, mentally and physically. Accordingly, this suggests we need to try to better control those waves to have better health—as well as improve our spirituality.

The Institute of HeartMath found that disharmony and negative emotions disorder the heart's electric waves, the responses of the autonomic nervous system and reduce immunity to disease. In contrast, positive emotions can harmonize heart waves and reduce stress to body systems as well as increase immunity.

We all know that stress can undermine our health. In stressful situations, the body's stress response system can become stuck in a partially 'on' mode, causing long-term damage and disrupting normal responses. This can include involuntarily adjusting heart rate, metabolism, immune system and other elements of the sympathetic and parasympathetic nervous system—which we know have an impact on the electric waves of our nervous system. This would obviously effect the energy systems that comprise our soul and spirituality.

A study regarding heart disease found one group of several groups of rabbits displayed 60% fewer symptoms of blocked arteries and associated symptoms, but for no apparent reason. That was, until the Ohio University study discovered the handler in charge of feeding the lucky rabbits petted each one for a few minutes before feeding. This human touch and affection was credited with reducing the affect of the toxic diet provided to all the rabbits. Repeated experiments, where one group of rabbits was similarly petted and another group was not produced similar results.

Another experiment found that love can make a big difference to our own lives. For example, children who grow up in homes where there is a lot of love and affection tend to be "better adjusted" psychologically than those who have the opposite, a childhood of neglect or abuse, and who display symptoms ranging from unhappiness to anger, to even greater neuroses.[9]

As Carol Rausch Albright notes, "Caretakers inform the infant who he is. If the caretaker communicates that he is wonderful, delightful, amazing, the infant returns eye contact and smiles in return. If the caretaker acts as if the baby were disgusting and burdensome, he apparently senses that. If no one pays much attention to the baby, many of his brain cells die off, and in extreme cases, so may the baby. Since the days of foundling homes, it has been observed that infants who are fed and cleaned without personal interaction are likely to die. In such places, an infant near a busy passageway is more likely to develop than a baby in a quiet corner, where no one stops by to offer some baby talk," she says.[10]

When young people believe their parents love them, they tend to grow into healthier adults, according to a long-term study by University of Arizona scientists. Students who rated their parents high for loving characteristics showed levels of disease far lower than those students who rated their parents low in terms of perceived level of love provided to them. "Love really matters," say researchers Gary Schwartz and Linda Russek.[11]

Similarly, Chopra reports a study with adult male victims of heart attacks which showed the most significant factor in their recovery was not anything to do with diet, exercise, smoking or a will to live. The men who lived felt that their wives loved them, while those who felt unloved tended not to survive. No other correlation the researchers could find was as strong.[12]

Various other studies have found that:

- depressed people who smoke appear to develop cancer at twice the rate of non-depressed smokers,
- introverted and depressed people develop more cancer and cardiovascular disease than their happier and more extroverted counterparts,
- of two groups of women with breast cancer, the group who received counseling lived twice as long as the control group that did not,
- of other women with breast cancer, those who had family support had greater immune activity than those who didn't have such support,
- breast cancer sufferers who thought positively about fighting it were found to have a stronger white blood cell count than those who thought they were "doomed" after diagnosis,
- when some cancer patients received counseling to offset depression, more natural killer-cells were found to be produced in their bodies,
- women with uterine cancer who reported more emotional stress than their peers at the time of diagnosis had a lower survival rate and shorter survival time,

- cancer patients who repress their spirit tend to have slower recovery rates than patients who are more extroverted,[13]
- single people are more likely to get cancer than people who are married,
- loneliness is considered a risk factor of cancer,
- a high level of perceived stress is a good predictor of lower natural killer-cell function and more upper-respiratory infections. (Killer cells fight viral and bacterial invaders), and
- people who have more stress tend to have more colds.

These are just a few of many studies that demonstrate the role of perceptions and beliefs in terms of health and recovery.

As we now know, perception involves electric currents in our nerves. And perceptions and feelings of love are known to help cohere electric waves in our minds and bodies.

These perceptions do not even have to be conscious perceptions. For decades, doctors assumed that an anesthetized patient was unconscious and therefore not aware of what happened to them in the operating room. However, it has since been discovered by hypnotizing patients after their operations that they were aware of what was happening during their operation, in particular what was said by doctors. If surgeons said a patient's prognosis was not good for recovery, patients tended not to recover. In contrast, those that unconsciously heard they would get better, tended to do just that. The more positive a surgeon's comment in the operating room the more positive outcome for the patient. So while it is supposed to be standard practice among doctors in surgery not to make negative comments, maybe they should proactively make positive comments about the outlook for the patient!

Joseph LeDoux adds that unconscious emotional memories of a trauma can also affect people who have do not have a conscious memory of it. The perception of the instrument or the sight of a person who inflicted the trauma can reactivate the emotional system and generate fear responses. These perceptions can even be generalized to other objects, and these "unconscious emotional memories can have widespread and long-lasting effects without a person having any understanding of what is triggering certain responses or feelings".[14]

Oschman in his book *Energy Healing: The Scientific Basis* proposes how this might work. He says it is likely that toxins and metabolic waste products accumulate in connective tissues, particularly in areas that have become desensitized as a response to trauma or structural imbalance. "The connective tissue gel can trap materials both mechanically (because of the small channels between its fibers) and electrically (because of its abundant negative charges)." He also suggests that applying the appropriate "pressure releases these trapped materials, some of which may have been stored for many years. They are released into the inerstitial fluid and carried awar by the lymphatic and venous drainage, and are excreted. At the same time, the application of pressure to the myofascial system producs piezoelectric fields and streaming potentials that stimulate the surrounding cells."

Candace Pert says takes this further and notes that healing can be blocked by an emotional state that has triggered the wrong chemicals. She says the immune system, like the central nervous system, has the ability to learn and remember. Intelligence is located not only in the brain but in the cells that are distributed throughout the body, and that the traditional separation of mental processes, including emotions, from the body is no longer valid. The body is the unconscious mind, she adds.[15]

While sciences, such as psychology, can help identify what causes such negative energy flows, they do not know how to convert them to positive energy flows. For example, Taoist Mantak Chia says "we have been taught that sexuality is sinful or dirty like garbage, and should therefore be suppressed. It is true that sexual desire is a major producer of negative emotions, but instead of repressing it, we can recycle and transform that same sexual energy into pure life-force energy," he suggests. "Although religion and society often try to suppress our sexual energy, the problem is that they offer no positive alternatives. They have not taught us how to recycle and transform this powerful force. When sexual energy is suppressed, it may end up expressing itself in other sense desires. Negative energies and emotions are not necessarily sinful, but they indicate that our life-force has been changed."[16]

Maybe the question is not "sex or no sex?", but rather what type of energy and perceptions are generated from it. The *Kabbalah*, for example, records "when sexual union is for the sake of heaven, there is nothing as holy or pure. The union of man and woman, when it is right, is the secret of civilization. Thereby, one becomes a partner with God in the act of Creation. This is the secret meaning of the saying of the sages: 'When a man unites with his wife in holiness, the divine presence is between them'....I once heard a chaste man bemoaning the fact that sexual union is inherently pleasurable. He preferred that there be no physical pleasure at all, so that he could unite with his wife solely to fulfill the command of his Creator: 'Be fruitful and multiply'. What he said helped me understand a saying of our rabbis: 'One should hallow oneself during sexual union'. I took this to mean that one should sanctify one's thought be eliminating any intention of feeling physical pleasure. One should bemoan the fact that pleasurable sensation is inherent to this fact. Sometime later, God favored me with a gift of grace, granting me understanding of the essence of sexual holiness. The holiness derives precisely from feeling the pleasure. This secret is wondrous, deep and awesome." In this case, while sex involves a lot of chemical, great sex can involved a lot of spirituality. (How do you tell the difference? Great sex prompts mental as well as physical orgasms).

Accordingly, how you perieve love-making is of great importance: do you sense the wonder of it, or rather feel guilty, or at worst unclean? The way you perceive and attach other emotions to this also coherent resonance of heart, brain and other body waves can influence your whole life, interactions and relations to and with other people.[17]

And this is not just restricted to this activity, but any activity you engage in. Accordingly, some elements of our health are in our own hands.

Yet, while many of us think we are what we eat, few of us pay particular attention to our spiritual diet. What we emotionally or spiritually chew upon, or what chews us up, is probaly just as important.

The *Tao Te Ching* and the *Bhagavad Gita* both offer similar suggestions as to how to improve the health of the soul. They suggest you close all the 'doors' of the body, shut the mind in the heart and channel your vital breath up through your head. For example, "in moving the vital breath [through the body, hold it deep and] thereby accumulate it. Having accumulated it, let it extend. When it extends, it goes downward. After it goes downward, it settles. Once it is settled, it becomes firm. Having become firm, it sprouts. After it sprouts, it grows. Once grown, then it withdraws. Having withdrawn, it becomes celestial. The celestial potency presses upward, the terrestrial potency presses downward. [He who] follows along lives; [he who] goes against it dies."[18]

In contrast, Buddhism suggests that energy moves from the brain to the throat, to the heart, to the navel to the genitals and vice versa.

Whatever the case, they indicate that we need to go with the spiritual flow of energy.

Besides general health, considering energy flows could also be important for treating ailments and accidents.

In her book *Infinite Mind*, Valerie Hunt refers to an incident in Kyoto in Japan in which she saw a woman with second degree burns to both arms receive a treatment that involved the wrapping of aluminum foil around the arms and connecting these with wire to acupuncture needles inserted just below the knees in the opposite leg. Hunt reports how the woman was free of pain in 20 minutes and the skin was normal after four daily treatments. The doctor explained that the treatment simply drained the excess energy from the burned arms.[19]

Electricity has been used in medicine, albeit vaguely, to treat peoples' ailments for centuries. One of the first recorded medicinal uses was in AD 46 when a Greek physician promoted the health benefits of standing on a wet beach near an electric eel.

Today, doctors measure electric signals via EEG and ECG to aid them in their diagnosis. Yet, many do still not recognize the importance of the body's electrical system. Just ask your local practitioner. However, knowledge is increasing. For example, research has shown that when a wound becomes inflammed it naturally enables increased electrical conduction. It is also known that certain low frequency electromagnetic fields activate the flow of calcium from cells. Calcium plays a key role in our cells and immunity, interacting with lymphocytes, a white blood cell that is one of the body's defenses against invaders.

Electromagnetism has also been used to directly treat problems of the mind and spirit, such as the infamous blunt instrument of electro-shock therapy. A refinement, "neuroelectric therapy" was used in Britain in the late 1970s onwards to treat drug addicts, including famous rock stars. This therapy used mild electric shocks to stimulate the production of naturally occuring painkillers in the brain, such as endorphins, which reduced both the patient's craving for drugs and helped reduce the symptoms associated with drug withdrawal. Endorphins attach to the same nerve cell receptor points as many mind-altering drugs, however, the former are natural and non-addictive. A person undergoing drug withdrawal has few endorphins in their body to control the pain. Neuroeclectric therapy stimulates their production. Some 80% of those treated remained drug free, according to pioneer of the therapy, Margaret Patteson.

In more recent uses, focused electromagnetism is being targeted to specific areas of the brain to treat epilepsy and Parkinson's disease. In terms of Parkinson's disease, a mild electric alters chaotic nerve signals and replace them with more coherent ones, says Canadian researcher Jonathan Dostrovsky.

This technique is also being tested for the treatment of pain, depression, obsessive-compulsive disorders, obesity and angina heart conditions. As the brain is an electrical organ, it makes sense to use electricity to modify its actions, says one of the doctors involved in this testing.[20] The theory is that appropriate electrical stimulation changes the abnormal pain signals, or corrects other electrical brain signals that are not functioning properly or won't shut down. The benefit of this approach compared to some earlier electric shock and brain surgery techniques is that it is easily reversible.

However, caution needs to be exercised for some other so-called energy therapies. For instance, a French composer claimed that by singing specific notes he could generate frequencies that would kill cancers.

The theory of being able to cure ailments by tuning into the "energy frequency" of the ailment and then correct it makes a lot of sense, except that there are not yet any devices or technology that can easily isolate and determine what those frequencies and patterns are and should be.

Though progress is being made, such as with magnetic resonance imaging and spectroscopy (MRI & MRS). MRI is based on the fact that nuclei of atoms in a cell line up like little bar magnets when placed inside a strong magnet. A second energy source is then applied, so they line up in a different manner. When this second source is removed, the nuclei return to their former state and emit a signal in doing so. This pattern is different for different types of cells and changes as cells become cancerous. MRS can distinguish between benign and malignant tissue with 95% accuracy

Another diagnostic device, an electromagnetic probe developed by Australian scientist Bevan Reid detects cancer and pre-cancerous cells by measuring the response of tissue to electric stimulation and comparing this to other computer-stored responses to determine if the tissue is cancerous.

It is forecast that assessment of different vibrating energy patterns of diseased and healthy tissues will be a core element of research later this century.

Interestingly, the Institute of HeartMath identified variability in heart wave rates as the most reflective and dynamic indicator of human emotional states. Accordingly, rather than just receiving a single simple ECG from your doctor, comparative ECGs could be a more effective measure of how you and your heart are faring.

Further research and new biotech developments are expected to see the market for electromagnetic-related medicine grow. However, energy treatments do not replace existing medical and pharmaceutical practices, but rather work at a supplementary level. Like pharmaceuticals, the quality and pattern of energy is very important, with the correct frequency and other parameters being crucial.[21]

20—The Power of Religion

Religion, like medicine, is similarly being challenged to interpret this new science of the soul and divine meaning from it. Religions have traditionally held the role of interpreter of creation stories to make sense of, and guide, our daily lives. However, for many people today, religious leaders are unable to adequately interpret recent tumultuous events.

Mythologist Joseph Campbell suggested one reason for this was that many religious organizations have tended to stress historical scriptures to the point they worship historical events instead of the spiritual messages within them, taking the stories and symbols literally rather than metaphorically.[1]

None-the-less, traditional religions provide some interesting insights into our souls and spirituality, especially given our new understanding of spiritual energy, which allows for further interpretation of religious metaphors. For example, spiritual practices from antiquity have dictated that we are unfinished as a person until we consent to a greater power and are drawn into a greater wholeness of being.

The religious symbolism of both Jesus and Orpheus is interesting in that they each leave the motherly earth to go to the heavenly realm of the father. If you take the mother and father anthropomorphic aspects away, you have an earthly energy being incorporated into a universal one.

The ancient Greeks and many Eastern religions also believe that divinity inhabits and is the very essence of us and the universe; it is not "out there" or "up above" as suggested in Western faiths.

The basic philosophy of ancient India's Rishis and Vedas is also that our real nature is a divine energy and that God exists in every living being.

Buddhism similarly suggests all things are infused with some level of spirituality and interconnected. And as physics shows us, vibrating strings of energy and the four fundamental forces connect everything in the universe. Tibetan Buddhist monks cite the spiritual as an active energy in nature, subtle but more powerful than the material. Buddha believed the human mind is capable of understanding everything, given enough natural ability, education and effort.

This sounds similar to a form of holographic field or even DNA, where one part can reveal the greater whole.

Interestingly, many spiritual leaders, such as Jesus and Saint Mother Teresa, are reported to have exuded positive energy; just being around them was considered beneficial to others.

These, and other, spiritual traditions say it is up to us to rediscover and reconnect with this divine power in and around us. But just what is this power? Is it the electromagnetic force involved in our souls and spirituality, or is it the as yet undetermined dark energy of the universe, or something else? And can further research actually tell us what this energy is and how to better relate to it?

This divine power does not have to be just viewed as a form of energy. If you want, and need to, still think of God or Allah as a man sitting on a throne somewhere, think of energy as the way this divine power impacts the world and you.

However, be aware of the limitations of imagining a finite image. For instance, the *Kabbalah* notes "an impoverished person thinks that God is an old man with white hair, sitting on a wondrous throne of fire that glitters with countless sparks, as the *Bible* states. Imagining this and similar fantasies, the fool corporealizes God. He falls into one of the traps that destroy faith. His awe of God is limited by his imagination. But if you are enlightened, you know God's oneness; you know that the divine is devoid of bodily categories—these can never be applied to God. Then you wonder, astonished: who am I? I am mustard seed in the middle of the sphere of the moon, which itself is a mustard seed within the next sphere. So it is with all the spheres—one inside the other—and all of them are a mustard see within the further expanses. Your awe is invigorated, the love in your soul expands."

Also, consider how the original Hebrew text of the *Bible*'s *Old Testament*, contains the word spirit or "ruah" more than 300 times. Ruah is a feminine word meaning breath. However, in the *New Testament* this word is translated to the neuter Greek word "pneuma", which also means breath. When the Romans translated the scriptures they used the masculine word "spiritus". This exemplifies how some definitions of spirituality can be distorted over time, complicating the challenge of accurately identifying the soul and its functioning. Similarly, consider the word "El Shaddai" as originally translated as "God the Almighty". Translators now suggest the word "Shaddai" may be divided into two parts, with "shad" meaning breast and "ai" being a feminine ending. The most literal translation of this phrase then is "God, the breasted one". Also, the *Dead Sea Scrolls* and the Nag Hammadi library of Egypt including the Gnostic Gospels refer to their believing in God as Mother, Father and Child, often referring to the Holy Spirit as the Mother. Interestingly, nature is still often referred to as "mother" nature.

Ex-monk now author Thomas Moore sums this up as most people being tempted to make God in their own image, rather than the other way around. "We feel the need for a certain source of security, and so we design a God tailor-made for our purposes. Then we defend the image against all others. But the purpose of the name of God is to crack us open, lifting us out of our finitude and self-absorbed anxiety, and this religious enterprise requires an idea of God beyond any fixed notion we might have," says Moore, in his beautifully written *The Soul's Religion*.[2]

As China's *Tao* says, to define God is not to have a true picture of God. It, like many other religious texts, says God is indefinable—the name that cannot be named.

This lack of a modern definition of God does not help people who are struggling to hold onto literal interpretations of ancient scriptures in light of recent events and scientific discoveries.

The *Kabbalah* suggests an answer. "All conceptual entanglements among human beings and all the inner, mental conflicts suffered by each individual result solely from our cloudy concept of the divine. Everything attributed to God other than the vastness of infinity is simply an explanation by which to attain the source of faith. One must draw a distinction between the essence of faith and its explanatory aids, as well as between the various levels of explanation. From learning and knowing too little, the mind becomes desolate, which leads to much thinking about the essence of God. The deeper one sinks into the stupidity of this mental insolence, the more one imagines that one is approaching the sublime knowledge of God, for which all the world's great spirits yearn. When this habit persists over many generations, numerous false nations are woven, leading to tragic consequences, until eventually the individual, stumbling in the darkness, loses material and spiritual vitality. The greatest impediment to the human spirit results from the fact that the conception of God is fixed in a particular form, due to childish habits and imagination. This is a spark of the defect of idolatry, of which we must always beware. All the troubles of the world, especially spiritual troubles such as impatience, hopelessness, and despair, derive from the failure to see the grandeur of God clearly." It adds "it is natural for each individual creature to be humble in the presence of God, to nullify itself in the presence of the whole—all the more so in the presence of the source of all being, which one senses as infinitely beyond the whole. There is no sadness or depression in this act, but rather delight and a feeling of being uplifted, a sense of inner power. But when is it natural? When the grandeur of God is well portrayed in the soul, with clear awareness, beyond any notion of divine essence. We avoid studying the true nature of the divine, and as a result, the concept of God has dimmed. The innermost point of the awareness of God has become so faint that the essence of God is conceived only as a stern power from whom you cannot escape, to whom you must subjugate yourself. If you submit yourself to the service of God on this empty basis, you gradually lose your radiance by constricting your consciousness. The divine splendor is plucked from your soul," the *Kabbalah* cautions.

Spirituality does not differentiate, only people and their churches do. Religion continues to be used as a political tool, a tool to keep people down and ignorant of their souls' potential, such as the claim by some religions that only they can "save" souls, that their devotees have to wear a certain head dress or uniform to be "saved". What does this mean for the souls of prehistoric people or those persons that were alive before a religion was founded? Surely they cannot simply be denied by God or Allah because they lived before the advent of a certain religion.

It is ironic that while most religions espouse the virtues of love and compassion (both Allah's and Jesus's main messages were that people need to learn to forgive and love one another) so much religious history is characterized by conflict.

While most of the world's different religions and spiritual institutions share the same basic tenets, there are obviously myriad ways that individuals experience religion.

Scientists suggest one reason for this is if a person has a spiritual experience (due to fear, thirst, jealousy or another trigger) and that the religious context says they have to kill other people who don't believe as they do, the result can be very dangerous. Researchers also note that while such people tend to intuitively know that what they may ultimately do (such as kills others) is wrong, the religious and cultural group dynamics overshadow their perceptions and individual spirituality. They compensate for a lack of personal spirituality by readily adopting someone else's definition of spirituality—i.e. a religion.

In short, the resonance of their soul is distorted by the emotional and other tags and perceptions provided by their culture. And without sufficient knowledge and emotional intelligence to counter this and let their true individual spirituality reign, their emotions enjoin the religious and cultural elements to the experience. In short, the harmonic wave patterns associated with their spirituality is distorted by the wave patterns generated by their perceptions, by their society and/or by their fears.

Maybe this is why Sigmund Freud suggested religion was an infantile neurosis. "God is the exalted father, and the longing for the father is the root of all religion." Conversely, Jung believed: "I am a splinter of the infinite deity..... I have never lost a sense of something that lives and endures underneath the eternal flux. What we see is the blossom, which passes. The rhizome remains." When asked if he believed in God, Jung is reported to have replied "I do not believe, I know!"

Interestingly, if there is only one divine power, as virtually all religions agree, why are there so many religions? Maybe the bumper sticker that proclaims "God is too big to fit in one religion" is correct. One explanation for why humanity expresses so wide a range of religious beliefs is that different people arrive at them from different directions, some from reasoned argument, others from blind faith, others from education, others from indoctrination, and yet others from personal and spiritual experience.

While they may provide spiritual guidance, religions do not have a monopoly on souls. Everyone who has electric circuits flowing through their nerves is a spiritual being. To describe someone as spiritual and someone else as not is simply to describe their differing awareness of and response to the universal search for self-transcendence, integration and identity—and ultimately their electromagnetic field.

Accordingly, being spiritual does not mean being religious, nor does being religious automatically mean you are spiritual. A spiritual experience cannot be ordained upon someone by a religion, but rather has to be experienced and interpreted individually, (or induced through Persinger's helmet).

As a result of such disparity, some churches have noted that they must reinterpret the scriptures in the context of contemporary culture, even science, if they are to be faithful to what was originally meant and help their constituents obtain spiritual enlightenment. Catholicism's Pope John Paul II, for instance, says *The Bible* "uses a symbolical language".[3] This is a little like reading Shakespeare; some people can understand the gist of the plots even when spoken in old English, while others need the settings and imagery to be more modern.

Religions could learn from this by combining the actual ancient allegory and what it means in today's modern world, using slightly more scientific imagery. The two are not always that far apart. For instance, religious scholars revisiting original documents and subsequent translations now believe the reason some scriptures said that God could not be named was not that it was forbidden to name the almighty, but that the words available and used then were limited in describing the all powerful. In another example, the common religious use of "universal breath" is just now beginning to be understood and described by physicists using the language of science, as in terms of string theory, cellular ion exchange and electric nerve circuits. Neither are wrong, rather they are very similar if you accept one definition as being allegorical for the other.

Another interesting issue facing religions today is should they evolve? Maybe they shouldn't evolve their fundamentals, though they could update how they communicate with people in today's modern world and take into account elements of society that were not applicable when the scriptures upon which they are based were written. For example, the various scriptures say nothing about driving motor vehicles, as cars were not invented then. Updating religion is not just a matter of delivering a 2000-year old message with 21st century pizzazz, despite today's marketing efforts of some churches.

Interestingly, Einstein believed religions already evolve through three stages:

1. the religion of fear and invention of dangerous gods,

2. the religion of a God who rewards and punishes, and

3. a "cosmic religious feeling".[4]

Some people add that this also appears to apply to individual human evolution of spirituality.

Einstein believed that religions, divested of their myths, basically do not differ from each other because the moral attitudes of a people, supported by religion, must aim at "preserving and promoting the sanity and vitality of the community".[5]

Despite some suggestions, Einstein was not an atheist, according to fellow physicist Max Jammer in his interesting book *Einstein and Religion*. In fact, Einstein described atheists as "they are creatures who, in their grudge against the traditional 'opium of the people', cannot hear the music of the spheres". In response to a question as to whether he believed in God, Einstein answered "I believe in Spinoza's God who reveals himself in the orderly harmony of what exists, not in a God who concerns himself with fates and actions of human beings." In another instance, he again indicated that a divine power created the world, when he said, "I wish to know how God created this world…the rest are details."[6]

Einstein summed up has spiritual belief as "I do not believe in the fear of life, in the fear of death, in blind faith. I cannot prove to you that there is a no personal God,…I do not believe in the God of theology who rewards good and punishes evil. My God created laws that take care of that. His universe is not ruled by wishful thinking, but by immutable laws."[7] In this, he suggested that God does not individually reward good or punish evil, but that there are laws of physics and nature that do this. Is this the different types of waves mentioned earlier, that provide us with a sense of what is good and bad, right and wrong?

Mathematician and philosopher Alfred Whitehead took this idea further and said that while God was responsible for ordering the world, this was not through direct action, but by providing various potentials which were able to be realized in the world and encouraging it towards a general goal—a kind of universal evolution.

More recently, physicist Paul Davies suggests there appears to be general organizing principles that govern the behavior of complex systems at higher organizational levels, principles that co-exist alongside the law of physics. "Through my scientific work I have come to believe more and more strongly that the physical universe is put together with an ingenuity so astonishing that I cannot accept it merely as a brute fact. There must, it seems to me, be a deeper level of explanation. Whether one wishes to call that deeper level "God" is a matter of taste and definition."[8]

This raises the question as to whether, and if so how, we can have a relationship with the energy of a divine power. Is it scientifically possible to have a relationship with energy?

Definitely. Energy flow can be a mutual process between you and the world around you. The laws of thermodynamics are quite clear about that, even if we don't know what that energy is.

Mediation and prayer would seem to be one way that the electric circuits of the brain, through coherence and stochastic resonance, might be able to better sense and somehow connect with an external energy.

Some studies have been undertaken in this respect. A study of patients with AIDS at the California-Pacific Medical Center in San Francisco in 1998 suggested those patients who unknowingly were prayed for by others became sick less often and recovered faster than those who were not prayed for. In a similar study at Duke University Medical School, patients who received prayers had up to 100 percent fewer side effects from cardiac surgery that those who were not prayed for. In contrast, a larger study by the Mayo Clinic found no difference between a group of coronary patients who were prayed for and those who were not.

Hunt suggests that for prayer to work it has to be on the same vibrational level as the universal energy of the creator force—if it is not to simply be a selfish wish. Further research is obviously required.

Whatever the case, if we are to commune with God we need to develop to God's level, not ours, otherwise we will never sense, let alone understand, what God says. God is unchanged, was the same yesterday, will be the same tomorrow. What has changed is our world and how we find and realize our soul and spirituality in it. Accordingly, do not expect God to change, rather you have to change yourself.

How do we find and realize our spirituality in today's world? Our new understanding of electromagnetism and fields opens whole new opportunities.

21—The Search for Meaning—Bringing it all Together

We each have a strong desire to understand our place in the universe, irrespective of your science or religion. It is a common human desire.

Today, many people want to know more about, and establish a better understanding with, God. Accordingly, we seek answers to such questions as, how does my life have meaning? How do I fit into the universe? What is the source of evil? What happens after I die?

Most cultures and their religions have developed a creation story of how the world began to help people answer such questions and make sense of their daily lives. The hard part is interpreting this information to reveal a common universal as well as individual truth that makes sense in today's world.

Unfortunately, science has not evolved enough to provide us with a theory of everything that could help answer this. And even when it does, such a theory is unlikely to provide a meaningful answer to what the meaning of life is.

However, some scientists have ventured to suggest a meaning. Davies says, "I have come to the point of view that mind—i.e., conscious awareness of the world—is not a meaningless and incidental quirk of nature, but an absolutely fundamental facet of reality. That is not to say that *we* are the purpose for which the universe exists. Far from it. I do, however, believe that we human beings are built into the scheme of things in a very basic way."[1]

Einstein described the search for meaning as "the individual feels the futility of human desires and aims and the sublimity and marvelous order which reveal themselves both in nature and in the world of thought. Individual existence impresses him as a sort of prison, and he wants to experience the universe as a significant whole." He suggested we can only find partial meaning. "He who knows nature knows God, not because Nature is God, but because the pursuit of science in studying Nature leads to religion…the manifestation of the divine in the universe is only partially comprehensible to the human intellect."[2] Accordingly, when looking for answers, besides looking inside yourself, nature can also be a good best place to turn as it reflects the laws of physics, energy and the creator force.

In contrast, Andrew Newberg simply suggests the brain is predisposed to having spiritual experiences, and that is why so many people believe in God.[3]

Others suggest spirituality and religion protect us against an exploding mystical experience that would be too much for us if we understood it. Matthew Alper suggests that with self-awareness the human species was transformed, or in biblical metaphor, man took the first bite of the Forbidden Fruit from the Tree of Knowledge. With this realization that we exist soon followed the ensuing realization that one day we will not exist and the anxiety of death and its associated reactions were realized. Our search for meaning is the result of this act of self-awareness.[4]

We can find other hints at meaning in various spiritual texts. For example, many spiritual and religious texts suggest we have lost our original nature and must rediscover it. Campbell noted that many myths commonly suggest: "You are God in your deepest identity. You are one with the transcendant....All of these symbols in mythology refer to you. Have you been reborn? Have you died to your animal nature and come to life?.... The birth of spiritual man out of animal man, it happens when you are awakened at the level of the heart to compassion and suffering with the other person, that is the beginning of hummanity."[5]

The *Kabbalah* says, "The purpose of the soul entering this body is to display her powers and actions in this world, for she needs an instrument. By descending to this world, she increases the flow of her power to guide the human being through the world. Thereby she perfects herself above and below, attaining higher states by being fulfilled in all dimensions. If she is not fulfilled both above and below, she is not complete. Before descending to this world, the soul is emanated from the mystery of the highest level. While in this world, she is completed and fulfilled by this lower world. Departing this world, she is filled with the fullness of all the worlds, the world above and the world below. At first, before descending to this world, the soul is imperfect, she is lacking something. By descending to this world, she is perfected in every dimension."[6] This suggests that developing the soul is a self-experiential or individual search process.

It is also similar in Taoism. "The human soul is a living ray of the cosmic soul," says Mantak Chia of the Universal Tao Center. "This is why we speak of the energy of human beings as both human and cosmic energy. The universe is the macrocosm, and humans are a microcosm of the universe. The more accurately we reflect the natural patterns of the universe, the more we are in harmony with the Tao. The more we are in harmony with the Tao, the more easily we can open to, unite with, absorb and enjoy the limitless energy of the universe."[7] This would suggest that we need to lead a "natural" life, whatever that may be. Taoists also suggest we need to learn to develop compassionate hearts and a wholeness of being; learn to help, heal and love others from the energies we receive from the forces of nature and the universe as well as learn about our own energy source and help release it within ourselves. "As we learn to control our life-force, we gradually develop the means to control our destinies," adds Chia. "There are two major approaches to spiritual disciplines. One may embrace concepts related to faith, good deeds, prayer right action, confession of sins and so forth. This approach gradually changes a person's negativity into virtuous energy, which is close to the Original Force or Wu Chi. Taoism offers another approach, involving the creation of a vehicle in the form of the energy body and eventually the spiritual body."[8]

Hinduism takes this further and indicates there are four basic types of people, with each type having different ways of experiencing God. These are:

- reflective people, advance toward God by knowing him;
- affective persons, who live more in their hearts than in their heads, draw close to God by loving him;
- active people, who advance toward God by serving him; and
- meditative people, who have an aptitude for meditation and realize God by this route.

Most people possess all four aptitudes to some extent.

After studying different religions, Campbell concluded God is a vehicle of energy, personification of energy. "The ultimate energy, that is the life of the universe. This is the mystery and it does not have to be personified, as in many Eastern, Asian religions it is more powers of nature. I see a diety as representing an energy system, and part of this is the human energy system. Angels and the like are metaphors for the energies that are affecting and guiding you."[9]

Today, physics and biology are showing what these Eastern religions have suggested, that the electromagnetic holographic paradigm can bridge the physical and transcendent worlds. For instance, we now know that we are fundamentally comprised of energy (vibrating strings of energy) and that our consciousness involves electromagnetic circuits, while God encompasses some form of universal energy. This reflects the ancient scriptures and modern science, though the latter is obviously not well understood at present.

The science of the soul also demonstrates that we no longer have to look beyond the world we inhabit to an after death heaven or nirvanic void to find our soul. It is all before us. Our nervous system processes information from the external world and at the same time realizes its own world in the process of cognition. It is no longer a question of nature versus nurture (the genes and body you are born with or the environment around you) but both. Nurture builds on nature. Accordingly, rather than awaiting death and judgment to realize our souls' potential, we are responsible for the development of our soul—now.

The science of the soul concurs with this; we can change our perceptions and alter our nerve circuits. Bishops, rabbis and clerics can show us paths and lead us in the right direction, but it is up to us to chart our own spiritual path.

Psychologist Carl Jung suggested self-knowledge and self-consciousness are the heart and essence of the search for meaning. "The psyche [a word Jung often transposed with soul] is transformed or developed by the relationship of the ego to the contents of the unconscious. In individual cases, that transformation can be read from dreams and fantasies. In collective life, it has left its deposit principally in the various religious systems and their changing symbols." He added: "Everything in the unconscious seeks outward manifestation, and the personality too desires to evolve out of its unconscious conditions and to experience itself as a whole. I cannot employ the language of science to trace this process of growth in myself, for I cannot experience myself as a scientific problem." Jung suggested our vision of ourselves "can only be expressed by way of myth. Myth is more individual and expresses life more precisely than does science....The only question is whether what I tell is my fable, my truth."

As for how to gain this self-knowledge, Jung noted "the decisive question for man is: is he related to something infinite or not? That is the telling question of his life. Only if we know that the thing which truly matters is the infinite can we avoid fixing our interests upon futilities, and upon all kinds of goals which are not of real importance. The more a man lays stress on false possession, and the less sensitivity he has for what is essential, the less satisfying his life. He feels limited because he has limited aims, and the result is envy and jealousy. If we understand and feel that here in this life we already have a link with the infinite, desires and attitudes change. In the final analysis, we count for something only because of the essential we embody. The greatest limitation for man is the 'self'; it is manifested in the experience: 'I am only that!' Only consciousness of our narrow confinement in the self forms the link to the limitlessness of the unconscious. In such awareness we experience ourselves concurrently as limited and eternal, as both the one and the other. In knowing ourselves to be unique in our personal combination—that is, ultimately limited—we posses also the capacity for becoming conscious of the infinite. But only then! As far as we can discern, the sole purpose of human existence is to kindle a light in the darkness of mere being."

He added, "I early arrived at the insight that when no answer comes from within to the problems and complexities of life, they ultimately mean very little. Outward circumstances are no substitute for inner experience. The years when I was pursuing my inner images were the most important in my life—in them everything essential was decided. It all began then; the latter details are only supplement and clarifications of the material that burst forth from the unconscious, and at first swamped me. It was the *prima materia* for a lifetime's work."[10]

Jung also made an interesting comparison between Eastern and Western approaches to spirituality. "The Indian's goal is not moral perfection, but the condition of *nirdvandva.* He wishes to free himself from nature; in keeping with this aim, he seeks in meditation the condition of imagelessness and emptiness. I, on the other hand, wish to persist in the state of lively contemplation of nature and of the psychic images. I want to be freed neither from human beings, nor from myself, nor from nature; for all these appear to me the greatest of miracles. Nature, the psyche, and life appear to me like divinity unfolded—and what more could I wish for? To me, the supreme meaning of Being can consist only in the fact that it is, not that it is not or is no longer. I cannot be liberated from anything that I do not posses, have not done or experienced. Real liberation becomes possible for me only when I have done all that I was able to do, when I have completely devoted myself to a thing and participated in it to the utmost. If I withdraw from participation, I am virtually amputating the corresponding part of my psyche. Naturally, there may be good reasons for my not immersing myself in a given experience. But then I am forced to confess my inability, and must know that I may have neglected to do something of vital importance. In this way, I make amends for the lack of a positive act by the clear knowledge of my incompetence. A man who has not passed through the inferno of his passions has never overcome them. They may dwell in the house next door, and at any moment a flame may dart out and set fire to his house. Whenever we give up, leave behind, and forget too much, there is always the dangers that the things we have neglected will return with added force. The world into which we are born is brutal and cruel, and at the same time of divine beauty. Which element we think outweighs the other, whether meaningless or meaning, is a matter of temperament. If meaninglessness were absolutely preponderant, the meaningless of life would vanish to an increasing degree with each step in our development. But that is—or seems to me—not the case. Probably, as in all metaphysical questions, both are true: life is—or has—meaning and meaningless. I cherish the anxious hope that meaning will preponderate and win the battle. When Lao-tzu says: 'All is clear, I am alone clouded,' he is expressing what I now feel in advance of old age."[11]

Another psychologist, Jean Piaget, demonstrated that human development of knowledge follows predictable steps, starting with children's physical exploration and progressing in adult's ability to think abstractly. Spiritual knowledge and experience is a further step in this progression. This suggests there is a set progression that we need to take in the development of our individual spirituality.

Harry Moody suggests there are five such phases in his *The Five Stages of the Soul: Charting the Spiritual Passages That Shape our Life*. These are what he categorises as: the call, the search, the struggle, the breakthrough, and the return, and describes the nature, pyschology and emotions that accompany each stage.

In his *Stages of Faith,* James Fowler suggests there are six sequential stages in the development of faith. These stages begin with protective faith in childhood to mythical-literal faith, through to conventional faith, reflective faith and conjunctive faith in midlife. He described the sixth and rarely achieved final stage as a progression to a universalizing life that is fulfilled by service to others. Fowler proposes that faith is an integral part of human personality and even those people who do not believe in God develop faith in something or someone. No one can escape confronting the meaning of their life and how they deal with this becomes a matter of their faith, he says.[12]

Whether there are distinct stages or continuous development of the soul, Hunt notes, "more people are discovering that the major job in life is attending to their own evolution. In every aspect, this is deeply spiritual. We have made strides in the physical, intellectual, and emotional improvement of our lot, but we have done precious little to manifest divine energy in our lives, our institutions, and our relationships. In short, we don't know how to be divine and human simultaneously. With spiritual maturity, life's primary directives come from the soul level, not from the material or intellectual level. In general, the course is unchartered; we have existed for so long emphasizing the human body-personality that we have little soul-self knowledge of our needs and weaknesses..... The divine part of our soul is indeed a rigid taskmaster, and when we stray from the path of spiritual evolution, it reminds us with unhappiness and with physical and emotional illness."[13]

As the Gnostic Gospel of Thomas says, the "Kingdom of the father is on the earth, but men do not see it". All the answers are here. You are just not looking in the right places.

Hunt suggests the meaning of human life "is to be both human and divine at the same time, which requires physical existence. Spiritual growth probably can occur when the soul is outside a physical body, but to achieve complete spiritual maturity, it appears that taking on human form is necessary."[15] She suggests that to hasten our spiritual development we should choose more carefully where, and on what, we spend our free time—as well as get more spiritual exercise.

While we each search for our own meaning and spirituality, there are some suggestions that there is also a collective purpose to our lives. For example, the *Kabbalah* says every human action on earth affects the divine realm. It is up to us to actualize the divine potential in the world. It adds that God is not a static being, that without human participation, God remains incomplete. In other words, God also needs us.

Physicist Max Jammer describes Charles Hartshorne's concept of temporalistic theism as where "God and the world condition each other, God's knowledge, which encompasses all realities, but only realities, grows and increase in an immanent way with the process of their development. God both participates in, and is enriched by, the processes of the natural order. God is not a transcendental being 'above' the world. Rather according to process theology, which identifies 'being' with 'becoming,' God is immanent in the world, just as the world is immanent in God."[14] In short, temporalistic theism and process theology suggest that God knows the probability of the actualization of a future event, but does not know in advance which of those events will be actualized, creating a cause and affect linkage between humanity and God.

This proposes that we don't just have an individual spiritual meaning to our lives, but also a collective spiritual mission. We are each specks of dust on a speck of dust floating among other specks. However, if you put these specks together you have something substantial. When we think of ourselves as centers of conscious or spiritual energy within an overall field of energy, we can begin to see that our lives are an integral and important part of a much larger whole—be it our family, community, planet or universe. Spirituality certainly seems to enable us to be more than the sum of our parts.

Is than a major purpose of our lives to unify our own energy fields with that of the universe, creator force or God; not just to be at one with the universe, but rather go a step further and help evolve a positive direction for mankind and the universe, and ultimately better understand and communicate with God? We, as a species, are hardly evolved spiritually. But as the one species to knowingly contemplate our spirituality, we are at least taking that first step.

Whatever the case, the journey we all take is important. As Eleanor Roosevelt once said, "The purpose of life, after all, is to live it, to taste experience to the utmost, to reach out eagerly and without fear for newer and richer experience."

The search for spirituality provides some meaning and personal significance for those engaging in it, to some extent providing a self-fulfilling prophecy.

Overcoming limitations and boundaries and developing new ways to engage novel experiences is one of the most important things we can do to develop our souls, as this allows us greater freedom to become aware of the world around us, of our perceptions, of our spirituality, and what we can become.

Proust took this a step further when he espoused "the real voyage of discovery consists not in seeking new landscapes, but in having new eyes."

Developing new ways of seeing the world implies this is not just a journey to gain new experiences that are incorporated into our electric currents, but that it is also a journey of identifying and understanding those circuits—and why we react the way we do. This enables each of us to embark upon a journey of better undestanding the universe around us, as well as that within.

Accordingly, we not only need to develop new ways of seeing, as we have done with this understanding of how physics is involved in our souls and spirituality; we also need to look at things, especially ourselves, through the eyes of others if we are to make the greatest possible advances in our spirit. By developing the ability to objectively look at ourselves from outside, as if through the eyes of an inpartial bystander, we can gain an understanding of how our electric waves and circuits are operating—and why the waves are the shape they are. This provides the power to see where we are in terms of Soul Power, where we need and want to go and what we have to do to take that journey.

If you seek to realize more spiritual experiences, the next section contains tools and techniques based on this electromagnetic understanding of spirituality.

PART THREE—PRACTICAL SOUL POWER: AND HOW TO GET IT

22—*Identifying Your Soul Power*

Spiritual development is an area where ancient knowledge, religion and the new age have tended to reign. But this does not mean that science is of no value to our spiritual development. Any thing that adds to our understanding can add to our spiritual progression. As the *Tao Te Ching* suggests, we develop our spirituality through understanding. "Understanding others is knowledge, understanding oneself is enlightenment." It adds, "To realize that you do not understand is a virtue; not to realize that you do not understand is a defect."

The *Bhagavad Gita* similarly says "the man unmoved when desire enters him attains a peace that eludes the man of many desires. When he renounces all desires and acts without craving, possessiveness, or individuality, he finds peace. This is the place of the infinite spirit; achieving it, one is freed from delusion; abiding in it even at the time of death, one finds the pure calm of infinity."

The Buddhist text *Dhammapada* extends this with "everything we are is the result of what we have thought". There is a similar Tibetan adage that says "don't wonder about your former lives; just look carefully at your present body. Don't wonder about your future lives; just look at your mind in the present!"

Tao philosophy adds a person can only reach heaven by first harmonizing their chi and their soul with the earth. It adds that the life force, or chi, should not be worshipped like a deity, but be seen as the power of God stepped down into our everyday world.

These and other scriptures are metaphorically analogous to modern scientific explanations of how we influence our own nerve synapses and circuits, and ultimately our soul and spirituality. For instance, LeDoux notes the way we think about ourselves can have powerful influences on us, the way we act and who we are. "One's self-image is self-perpetuating." He also says, "that the self is synaptic can be a curse—it doesn't take much to break it apart. But it is also a blessing, as there are always new connections waiting to be made. Your are your synapses. They are who you are." He adds: "In my view, the self is the totality of what an organism is physically, biologically, psychologically, socially and culturally.

Though it is a unit, it is not unitary. It includes things that we know and things that we do not know, things that others know about us that we do not realize. It includes features that we express and hide, and some that we simply don't call upon. It includes what we would like to be as well as what we hope we never become."[1]

Accordingly, while the universe may have been created just once, we can recreate ourselves with each experience and thought. This process of "recreating" our perceptions, our nerve circuits and ultimately ourselves provides us with the power to influence our souls and even direct our spirituality—Soul Power. However, before you can enact Soul Power you need to consider the current state of your spirituality.

It is an interesting fact of life, that many of us undertake various courses and training to get better at what we do, but we don't do the same with our soul. For example, while many people use plans to construct houses and have financial, marketing and business plans, very few have plans for their life and spirituality. There is only one letter difference between goal and God. Why not start one? (A subjective, yet practical way to do this is included in the footnotes.[2])

While we can't control our lives just by thinking, we do have choices that we can make, choices that influence the electromagnetic waves in our brain and body which can in turn influence our perception, learning and ultimately our soul. The same events happen, we just modify how we perceive and respond to them. It is not mind over matter, rather that mind and spirit matters—and being aware of this.

Accordingly, recognizing the perceptions and decisions we are now making is one of the best ways to take possession and control of our spiritual lives. Many new age books have been written on this topic, ranging from "say these affirmations and your life will change" to "take more control of your life" approaches and so on. Books, such as those by Deepak Chopra, Thomas Moore, Gary Zukav and many others provide suggestions on how to care for your soul.

Given the new objective understanding of how the soul and spirituality operate, there are several things that we can do to improve our spiritual development. However, the fact that we have somewhat explained the "hard-wiring" of our spirituality does not guaranteee that we will be spiritual, we each need to switch it on and develop it.

23—*How is Your Soul Now?*

It is not always easy to switch on, and take control of, our spirit. Our nerve circuits and emotional tags unconciously make many choices for us.

Accordingly, understanding what we are perceiving, thinking and feeling is important. Otherwise our nerves and emotions will do it for us unconsciously. As LeDoux notes, if we do not know what it is that we are learning about, those stimuli and perceptions might later trigger fear responses that will be difficult to understand, even difficult to control.[1]

Fear is one of the biggest obstacles to spiritual transformation. Psychologists note that fear dimishes creativity and isolates us. When the soul faces too much fear, it appears to begin to shut down and go numb. This fear can be of the unknown or simply fear of what your friends will think of you if you express your spirituality. You are not alone. Some of the best marketing techniques work by playing on your fears, such as your cannot survive without this product or service and your life will suffer as a result. Studies have proven that the fear of loss approach is more effective in marketing than appealing to desire or greed. Nobel Prize winners Daniel Kahneman and Amos Tversky in their Prospect Theory found that people care twice as much about avoiding losses as they do about making gains. This not only applies to psychology, finance and marketing, but also to how we try to grow, as most of us are concerned about "losing" some aspects of ourselves.

To free yourself from fear and other negative emotions, you need to become emotionally aware, or emotionally intelligent. Without being emotionally aware you can be hostage to your emotional compulsions and the emotional tags associated with your perceptions and thoughts.

This means not just knowing when you are frightened of angry, but also determining the underlying reason why you are scared, angry or emotional. If you can do this, you can determine how to change the emotional tag that has been associated with a perception or memory—and ultimately free your spirit.

For example, when you feel your heart racing, your palms sweat or fear arise, review the situation. Is the situation worth the jagged electric waves that it is creating inside you?

Whatever the situation, often you can take some control over how you perceive it; such as viewing something as a terrible situation but realizing it is making you a stronger person, or the fact that there is always someone worse off. While influencing the perception does not change it, this can be enough to help you better address it, or at least not attach a damaging emotional tag to it.

As the Institute of HeartMath notes, "Thoughts and emotional states can be considered 'coherent' or 'incoherent.' We describe positive emotions such as love or appreciation as coherent states, whereas negative feelings such as anger, anxiety or frustration are examples of incoherent states....different emotions lead to measurably different degrees of coherence in the oscillatory rhythms generated by the body's systems..... The term coherence is also used in mathematics to describe the ordered or constructive distribution of the power content within a *single* waveform. In this case, the more stable the frequency and shape of the waveform, the higher the coherence."

It says, "The key to the successful integration of the mind and emotions lies in increasing the coherence (ordered, harmonious function) in both systems and bringing them into phase with one another. It is our experience that the degree of coherence between the mind and emotions can vary considerably. When they are out-of-phase, overall awareness is reduced. Conversely, when they are in-phase, awareness is expanded. This interaction affects us on a number of levels: vision, listening abilities, reaction times, mental clarity, feeling states and sensitivities are all influenced by the degree of mental and emotional coherence experienced at any given moment," the institute reports on its website.[2]

This brings us to another key element in developing Soul Power, it must be positive in order for it to achieve a positive result. Negative imagery will achieve negative results.

Positive thinking, positive beliefs and feelings are the basic foundations for peace and joy, says Tibetan lama Tulku Thondup.

This is supported by scientific studies that show most people prefer repeating a positive experience than a negative one. A study by the National Dental Network found patients most liked warm, friendly and open behavior from their dentist. The study found that patients did not so much care to understand the clinical skills of dentists, but rather wanted to get to know their dentist's approach, their passion for their work and their level of concern for their patients. Interestingly, many dentists felt that they needed to let patients know a limited amount about their own personal life in order to establish a level of mutual trust.

In another study, the Center for Creative Leadership found a common trait of good leadership is optimism. Thinking positively is not only good for your personal life, but also your career. One reason given for this is that optimism tends to encourage people to continue in the face of adversity. The center also notes that it is possible to learn optimism and positiveness.

Other research found a positive attitude is also important for sports stars. Researchers at the University of Toronto found the most vital factor in making the jump from amateur to the National Hockey League is optimism. A player who gets angry and blames others after a loss takes on average another three games to recover compared to a player who is optimistic.

If that is not enough reason to be more positive, recall the study that demonstrated having a positive self perception can also increase your life expectancy. A positive attitude about aging can extend a person's life by more than seven years, say researchers at Miami University. The way a person perceives aging outweighs gains in lifespan from qutting smoking or exercising regularly, found Suzanne Kunkel, director of the university's Scripps Gerontology Center. The study found respondents with more positive views on aging lived longer, even after taking into account factors such as socio-economic status, health and loneliness among others. Those with more positive self-perceptions lived a median of seven and a half years longer than those who had more negative perceptions.

Further research found one of the main predictors of longevity is a persons' ability to incorporate the negative affects of ageing without actually feeling unwell. Harvard University psychiatrist George Valliant found people who could do this tended to expect the ailments of aging and illhealth to pass, did not feel like victims, continued to plan and hope, also displayed empathy and gratitude.

Research into happiness by psychologists also provides some other interesting insights. To be happy, we need to train ourselves not to make a big deal out of trivial hassles, but rather learn to focus on the process of working towards our goals and not wait to be happy until we achieve them, says Ed Diener.[3] Also, make a habit of noticing the good things in our lives. To be happier, he also suggests we need good friends and family, people who care about us and about whom we care deeply. Diener adds we should involve ourselves in activities that we enjoy and value. Materialism is toxic for happiness, he notes; even rich materialists aren't as happy as they would like.

Similarly, psychologist David Myers suggests that to be happy we need to give priority to close relationships. "There are few better remedies for unhappiness than an intimate friendship with someone who cares deeply about you. Confiding is good for soul and body," he writes in *The Pursuit of Happiness.*[4] He adds, "Focus beyond the self. Reach out to those in need. Happiness increases helpfulness—those who feel good, do good. But doing good also makes one feel good.

Compassionate acts help one feel better about oneself." Myers also advises, "Act happy. Talk as if you feel positive self-esteem, optimistic and outgoing. Going through the motions can trigger the emotions."

The happiest people spend the least time alone, pursue personal growth and intimacy, and judge themselves by their own measures not against what others do or have, says Martin Seligman of the University of Pennsylvania.[5] People also have "signature strengths," and the happiest people use them. He also notes that optimistic people tend to minimize any causes of misfortune and distance themselves from them. In contrast, pessimists blame themselves for their misfortune and attribute good things to chance rather than themselves.

Forgiveness is the trait most strongly linked to happiness, according to psychologist Christopher Peterson. "It's the queen of all virtues, and probably the hardest to come by."[6] Dale Carnegie, author of *How to Win Friends and Influence People* says, "When you are good to others, you are best to yourself."[7] Acceptance and compassion are also regularly cited as an important element of spirituality.

Others suggest that simply smiling can help in terms of increasing positiveness and happiness. A warm smile relaxes the recipient and often sees them let down their defenses, scientists have found. This "relaxation response" is caused by physiological changes in the nervous system, with the respiratory system, circulation, muscles and endocrine system all impacted by receiving a smile. Giving a smile raises the chances of receiving one.

You may also want to try to laugh more often: laughter also changes body chemistry, which can make it easier to change the mind and spirit.

This and other research shows how feeling positive, being caressed or hugged, receiving a massage, loving, or even just thinking about these can have a positive effect on our physiology—ranging from a lower heart rate, reducing stress, to increasing immunity and improving our ability to resist or recover from illness.[8] Also recall those studies mentioned earlier that show how happier and extroverted people develop less cancer and cardiovascular disease compared to their unhappy and introverted counterparts, how breast cancer sufferers who thought positively about fighting it were found to have stronger white blood cell counts compared to those who thought they were "doomed" after diagnosis.

When people are happy and in positive moods the most active part of the brain are nerve circuits in the left prefrontal cortex, in particular the left middle frontal gyrus, an area just above your left eyebrow. In contrast, when people are emotionally distressed, angry, anxious or depressed, MRI images and EEG analysis reveall the most active areas are circuits converging on the amygdala and the right prefrontal cortex, notes psychologisy Richard Davidson.

Davidson can even determine a person's range of emotional moods by measuring the ratio of electrical activity between the left and right prefrontal cortex. The more happy and positive a person is the more the ratio tilts to the left. In contrast, those people who are unhappy or distressed have a ratio that tilts to the right. The greatest reading recorded on the positive left was from a Tibetan Buddhist monk, he notes. In another instance, when another monk was asked to meditate and generate a state of compassion, the electrical activity in the left prefrontal cortex increased 800%.[9]

Davidson also notes that our emotional range can be shifted, with appropriate mental training. Such training strengthens neurons in the left prefrontal cortex and possibly inhibits electrical messages from the amygdala that drive opposing emotions, he theorizes. Another benefit of shifting emotional mindfulness towards the positive, he found, is that it also improves the robustness of a persons' immune system.[10]

Aside from the statistics, research and improved health, think of the practical applications of being more positive and happy. For instance, when it comes to dating or finding a mate, do you want someone who is positive and builds you up or pulls you down? This also applies to the way you deal with people, be they acquaintances, friends, colleagues or superiors. Most people that you deal with in life and work want to deal with positive people and prefer repeating a positive experience than a negative one. "It is our basic right to be a happy person, happy family, and eventually a happy world," says the Dalai Lama. "That is all right. That should be our goal."[11]

However, being positive, happy and spiritual is harder, much harder, than being negative. That is why so many people are so negative about so many things. Negative emotions tend to initially be stronger; though love is ultimately more powerful.

But being positive and going around smiling isn't the whole approach to Soul Power. Being positive also means being proactive. This means creating and taking advantage of opportunities, not waiting for opportunities to be brought to you. This increases the probability of capturing good opportunities, though it does not ensure that you will capture them. That is up to you and your own personal skills.

In fact, research indicates that we can no longer blame our parents or our genes for our negative thoughts and traits. Our genes account for less than half of any particular personality trait, states LeDoux.

LeDoux notes in an a media interview that "life's experiences, in the form of learning and memory, shape how one's genotype is expressed. Even the most ardent proponents of genetic determination of behaviour admit that genes and environment interact to shape trait expression. While the fact that both nature and nurture contribue to who we are is widely acknowledged, less well understood is that, from the point of view of how the brain works, nature and nurture are not different things but different ways of doing the same thing: wiring synapses. That is, both genes and experiences have their effects on our minds and behavioural reactions by shaping the way synapases are formed. "Through learning and memory processes, and the underlying synaptic changes, personality builds up in a cohesive way. Without learning and memory, personality would be an empty expression of our genetic constitution. Learning allows us to transcend our genes," he says.

"Synaptic interactions between the systems that underlie the individual processes are key in keeping the self integrated in space (across brain systems) and time (across the days of our lives)," he adds. "Synapses are simply the brain's way of receiving, storing and retrieving our personalities, as determined by all the psychological, cultural and genetic factors," he says. Interestingly, LeDoux also notes how synaptic connections are also key to many mental disorders. "These were long thought of simply as chemical imbalances. In fact, the key is not the chemicals themselves, but the circuits in which the chemicals act," he says.[12]

Accordingly, while genetic and environmental influences are very powerful, they do not control us. We are not the product of our past, we are not victims. We have some control of how we shape our spirit. We become what we think about, and our attitudes about our performance can be self-determinative, adds Richard Restak, professor of neurology and author.

In short, altered traits are more important than altered states. At first, they may not seem as spiritual as altered states of consciousness, but ultimately they are in terms of cohering your body and brain waves—and spirituality.

As the Persian poet Rumi suggested, "May God cause you to change your life in the way your know you should. He has so completely dissolved his ego, nothinged himself, that what he says is like God talking to God. The love-religion has no code or doctrine."

And as the *Bhagava Gita* adds "know your self, for your self the consciousness is all in all."

The way your electric circuits and waves perceive and react to things is the deciding factor in causing happiness or suffering. If the electric waves throughout your mind and body are chaotic you will not be happy even in the best of conditions. In contrast, if these waves are harmonious, you are more likely to remain serene even in adverse environments.

Changing these synapses and electric nerve circuits that ultimately form our perceptions, our character and our spirit is not always easy. But it can be done. Discipline is required to make the choice, over and over again, to interrupt the cycle of negative thoughts and replace them with thinking and feeling positively, being happy, loving, compassionate and other virtues extolled by various scriptures that can smooth, cohere and strengthen those electric waves. And the more we react positively, the more it leads to increased positive feelings and expression by our nervous system and our soul.

24—Growing and Exercising Your Soul

One of the ways we can influence our spirituality is through creating more harmonious brain and heart waves. If the waves generated by the electric currents in our nerves are harmoniously smooth and coherent, as generated during happy and positive states, there is less stress on the system.

With each beat or wave the heart continuously communicates with and influences the brain and other bodily system, notes the Institute of HeartMath. The institute has demonstrated "the electric waves messages the heart sends to the brain not only affect physiological regulation, but also profoundly influence perception, emotions, behaviors, performance and health.

"In turn, the heart's output is measurably influenced by our-moment-to-moment emotional experience." The institute says, "Our research shows that negative emotions lead to increased disorder in the autonomic nervous system and the heart's rhythms, adversely affecting the rest of the body. In contrast, positive emotions increase harmony, synchronization and balance in the nervous system, which is reflected in more 'coherent' heart rhythms.

"Even more intriguing are the dramatic positive changes that occur when people use techniques to intentionally experience positive emotions and increase coherence in the heart's rhythms. Not only does a beneficial cascade of neural and hormonal events begin, but profound shifts in perception can also occur, and clarity of mind increases. We gain greater emotional stability and become more able to effectively reduce stress and anxiety, while improving learning and performance."

The institute points out "in sum, from our current understanding of the elaborate feedback networks between the brain, the heart and the mental and emotional systems, it becomes clear that the age-old struggle between intellect and emotion will not be resolved by the mind gaining dominance over the emotions, but rather by increasing the harmonious balance between the two systems—a synthesis that provides greater access to our full range of intelligence."[1]

Our nervous system, in particular the sympathetic and parasympathetic nervous system, continually and often unconsciouslly reconciles and balances stimuli and information from outside with what is already stored in our nerves circuits. Immaterial electromagnetic energy is combined with impartial atoms, nerve cells and chemicals to create such subjective things as human consciousness, emotions, character and spirituality. Energy is enjoined together to accomplish a transcendent task that no single element could do on its own: our souls and spirituality are much larger than the sum of their components.

Using this objective understanding of the electromagnetic waves and framework of our souls, we can further develop our spirituality. This starts with being aware of long-term potentiation (LTP) and the role electric circuits and wave principles, such as coherence and stochastic resonance, play in forming perceptions, learning, memory and ultimately spiritual experiences.[2]

Applying techniques to harmoniously cohere brain and heart waves and reducing those stressful jagged waves should improve the way your perceive events as well as reduce stress and improve your health, mind and ultimately your spirituality.

Recalling and invoking memories, for example, can result in the replaying of the energy pattern associated with that memory. Memories and thoughts of love can cause the heart to beat faster and waves associated with this feeling to be conveyed throughout the body. Or in the case of a fear, more jagged waves of energy can cause your skin to sweat. This is interesting in that while the initial stimulus may no longer be there, the mind can cause the body to undergo a similar response as if it were.

This recall provides a powerful tool that allows us to use positive thoughts to generate harmonious waves and flows of energy that can counter negative thoughts and flows—and influence our own minds and souls.

We can then also use the laws of physics, in particular the laws regarding energy, electromagnetism and waves to influence and modify our brain and body waves. This should ultimately help raise our consciousness and grow our souls and spirituality—and maybe increase the probability of having more spiritual experiences.

The first law of thermodynamics, for example, states that energy can not be created or destroyed. And you can't divorce energy from your soul anymore than you can separate energy from matter. Accordingly, the energy in your nerve circuits has to keep flowing and circulating. Electric waves are going to flow in your head and heart no matter what you do.

You can't stop electric flows without creating blockages and eventually harming yourself.

Letting energy flow sometimes requires letting go of old or odd electric circuits/thought patterns. As such, don't let family issues, religious or other conflicts choke-up inside you. Don't live in the past, in what was; rather live for what can be. Psychologist and author Tim Kasser, also says people who emphasize the pursuit of money, possessions, personal appearance or fame and popularity report lower psychological wellbeing than people with less materialistic values, those who emphasize self-acceptance and personal growth, intimacy and friendship.[3] Analyze any negativity, such as why do negative or violent feelings arise? What is your energy doing that results in you being angry? Are you trapped in a circumstance that you cannot escape from, with the parasympathetic flight response overwhelmed by work or social commitments and creating jagged electromagnetic waves? Or is it just an emotional tag mistakenly applied to a perception or situation, such as fear of not knowing how to perceive or respond being channeled into anger?

Once you better understand the situation the better able you are to channel energy flows to a more constructive and positive response. It is not a matter of whether you experience stress or not, says Marilyn Albert of Harvard Medical School, it is your attitude toward it that is important.[4] Remember, it is how we channel energy flow that creates LTP and ultimately defines ourselves.

If you are feeling spiritually empty, do not despair. Such emptiness frees us from the anxiety of having to be in control and gives hope, suggests monk turned psychologist and author, Thomas Moore. He notes that early in his own psychological consulting practice he discovered people coming to him for therapy had little faith in themselves. They were overwhelmed by troubles that stole their attention and they could not see their own worth, though he could see their strengths and tried to show he had faith in them. Other psychologists, theologians and new age practitioners suggest when you are out of touch with your spirituality that you can turn on spirit by simply behaving more spiritually.

This brings us to another law of physics that is important for us, momentum. Momentum persists unless it is changed by a further force, and this is true of nerve circuits and waves. Energy will tend to flow along an established path unless there is another force to create a new direction. For example, a wave in a pond will not change direction until it intersects another wave. Again, it is up to us to apply further energy to form new paths and directions.

There are several tools and techniques to do this. One is where attention goes energy flows. If you focus your attention on something, such as improving your

Soul Power, your energy should similarly flow toward this task. "Not possible," you say. Think again.

It has been proven that people can alter the electromagnetic circuitry of their brains to a healthier pattern. Patients can learn to weaken the physical connection to the old pathological scheme and strengthen those to a new, healthier one, notes neuroscientist Jeffrey Schwartz and Sharon Begley in *The Mind and The Brain: Neuroplasticity and the Power of Mental Force.*[5] They note how a group of subjects practiced a piano piece and another group who just thought about practicing the piece, playing each note in their head without actually touching a piano, both showed physical changes in their brains. This is analogous to athletes imaging an event in their minds before they do it.

Schwartz also found that, with appropriate training and actively focusing their attention away from negative thoughts and behaviors towards positive ones, patients were able to affect significant and lasting changes in their own neural pathways—and perceptions and actions. It is not a case of mind over matter, but rather that the electric mind changes brain matter and all the associated chemistry. Schwartz and Begley identified and proved that people can use their own minds to reshape their own minds.

And with the ability to shape our brains and minds comes the power to shape our destiny. It takes just a small step further to understand that we can also shape our spirituality!

What tools and techniques can you use to reshape those electric circuits in your own head; how can you harness the principles of waves and electromagnetism to improve your Soul Power? There are several that have been proven to be affective.

Prayer, for example, is one tool that is known to be able to focus attention. Science has now shown what the religious have known intuitively for centuries. Focusing on and repeating a prayer can influence brain and heart wave patterns.[6] Some people suggest that when prayers are offered in the "right" framework (whatever that may be) they provide a direct connection to God and the universe, while others suggest that for prayer to be effective it has to be in resonance with universal energy, otherwise it is just a wish. Interestingly, people who pray daily are twice as likely to say they are happy compared to those who do not, found one report.[7]

Various religious rituals and ceremonies can also focus attention. And maybe the community or group environment aspect can also help cohere thought waves.

Meditation can similarly help focus attention. Interestingly, meditation has tended to focus on controlling body energy and waves via breathing. Studies have found that meditating about love, compassion and caring generate greater coherence in your heart waves (ECG) than people who are simply resting. Interestingly, certain meditations enable practitioners to choose whether they will experience an emotion or not, and even what emotion they will experience—they can choose and control which emotions they feel.[8]

Reading motivational, spiritual or new age books can also be beneficial, by promoting coherent energy flows within the reader who takes the anecdotes and stories to heart—or should I say changes their heart and brain waves to a more coherent pattern. With our understanding of physics and spirituality, we can also better determine how some of the miraculous events, stories and cures cited may have come about, and what the actual physics are that are often invoked but not explained.[9]

Other techniques to change energy flows include Jin Shin Jyutsu and yoga among many other physical actions. Self-help Jin Shin Jyutsu is said to be able to jump start energy flows and overcome blockages within the body realign energy patterns. It appears to be able to do so by invoking the principal of wave momentum, whereby the electric waves of the practitioner are added to and alter those of the subject. If the energy field of one person is similar to that of another being treated in terms of frequency, patterns and so on, touch could provide more energy to the subject, just as parallel electric circuits offer more power than series ones. As noted earlier, various studies, have shown that touch can influence our own energy flows.[10]

Mudras, or finger yoga positions, can also be useful for focusing energy. Posing the fingers in certain positions are claimed to aid the flow of energy throughout the body.[11] Buddhas, for example, are depicted with various finger and hand positions that provides benefits ranging from increased serenity to greater intuition. Mudras appear to work by prompting those practising them to stop and take time out and to focus attention—and where attention flows energy goes, leading to changes in the flow and patterns of electric waves. While Eastern religion has practiced Mudras for thousands of years, modern science has yet to investigate the basis of them.[12]

Some people suggest another way to channel energy flows is to spend time in nature and absorb natural energy flows. For example, Jung suggested in his biography that "it seemed to me that the high mountains, the rivers, lakes, trees, flowers, and animals far better exemplified the essence of God than men with their ridiculous cloths, their meanness, vanity, mendacity, and abhorrent egotism."[13] Similarly, Einstein wrote: "Our task must be to free ourselves…by widening our circle of compassion to embrace all living creates and the whole of nature and its beauty."[14]

Others suggest just letting energy flow and see where it goes and what thoughts and feelings it generates. Some people ascribe the free flow of energy to be the source of creativity, self-expression and passion. However, remember to be aware of what you intuitively know is right and wrong as you do this. Interestingly, some new age practitioners describe creating a wave of energy and riding it. Most of us can understand riding a wave of water as in rafting or surfing, even riding a wave of wind in a glider or even a wave of gravity in skiing, but learning how to do this with our minds and spirit is a little harder to get an objective mind around.

More mechanical tools, such as a pendulum clock with a slow tic-toc sound, can also be a good way to help cohere wave rhythms in your space. Similarly so can the sound waves of music. Using the right rhythm for you can harness the princples of wave interaction to create coherence and positive change.

Then there are various chemicals and drugs that can influence the flow of energy. For example, antidepressents have been shown to be able to alter nerve circuits.[15] However, one third of patients treated with antidepressants do not respond to them and of those that do, only 50% get completely better, suggesting that it takes more than just chemicals to provide the total cure to depression. Patterns of positive energy flows can explain how people who are more positive about their illness recover quicker than those who are not as positive. Remember, how coherent waves are known to improve immunology and other bodily functions, as shown by the Institute of HeartMath and other researchers.

All the above-mentioned tools and techniques ultimately harness the principles of wave interaction, coherence and stochastic resonance, to strengthen or cancel waves.

If you can get your brain and heart waves to resonate more coherenetly and join together to form stronger waves you should be able to energize yourself and elevate your spirit. Taking this simplistic analogy further, a vessel that sails against the waves can be unstable and even capsize, whereas a captain who knows how to sail with the flow will get where he or she wants to go more efficiently.

The wave effect also explains why stress is so debilitating. Stress involves the parasympathetic nervous system countering the actions of sympathetic nerves. One set of electric waves counter the messages of the other, resulting in a jagged wave pattern that affects our brain, heart and ultimately our body, mind and spirit.

This can also explain why stressed people can get tired without exercising. If you are continually worrying or in a state of nervous conflict, your nervous system could be using a substantial amount of energy, but the waves cancel each other out leaving you feeling drained.

Similarly, if we find ourselves doing something that resolves the level of conflict between the sympathetic and parasympathetic nervous system by redirecting the energy flow, such as eating, that might become the preferred way to deal with a stressful situation. This could explain why some of us eat more when stressed. And in turn, the more stressed we are the more we eat to channel our energy towards that task Remember how the stomach contains its own nervous system capable of learning and memory—and how it is capable of generating its own electric waves to communicate with the brain and the rest of the body. Eating would pre-occupy those nerves, and if they were emitting jagged waves, digestion could reduce their impact on the rest of the nervous system, making us feel slightly better.

Add more negative waves from other sources or emotions and the waves can become even more chaotic and perceptions and memories distorted negatively. This can manifest itself in a range of ways, such as increased fear or phobias. A "negative" stochastic resonance may also be invoked and make things worse—or give rise to what some people call evil.

To decrease energy, we have to send it somewhere. Again, this is best done by channeling it into something constructive. It is up to us to find something positive, something constructive to reform those jagged waves. This is one reason why exercise is so popular, as it channels and consumes energy.

Another important scientific principle is Newton's law that for every action there is an opposite reaction. For example, a wave in a bath will reach one end and then bounce back again. (Science describes this principle in terms of an equal and opposite reaction, but with a multitude of factors involved in modern life this is not always apparent.)

This law not only applies to physical systems such as the human body, but also to the mind and spirituality—where you get back what you put in. For instance, if you put in someone else's dogma, that is what you get back; whereas if you put your own effort, you will get something more individual back. Various scriptures, as well as studies mentioned above, suggest that putting out "bad" energy results in bad energy flowing back, while putting out good energy and thoughts results in this being returned in due course. This could explain the metaphoric parables of religious scriptures, such as "what you sow is what you reap" or "do unto others as you would have them do unto you" and is analogous with the new age's karma of what goes around comes around. (It is interesting to go back and reread scriptures in light of energy-based spirituality and realizing how much sense the metaphors now make.)

Similarly, you cannot cross the street without making some effort. Once you decide to cross an unknown road, it is then up to you as to how you cross it. Do you seek a map, run blindly across and hope to hit nothing, watch what happens for a while then cross, wait for a sign, or wait for someone who has crossed before to assist you. There are many ways, each providing a different experience but still getting you to the other side. In contrast, if you just lie back and wait it is likely that not much is going to change in your life the way you might want it to.

Another important point about this law of physics is that it also provides a basis for how exercising our minds can make them grow. Recall how we either use our nerve circuits or lose them. New experiences ultimately generate new nerve circuits, with brain scans showing that we use a lot more of our gray matter when undertaking new or strange tasks than when we are doing familiar ones. Another study showed that rats raised in bright cages with toys to play with grew more neural connections than rodents raised in bare cages. This could explain why travel and new experiences in new locations are so desired. In contrast, other research has shown that stress produces fewer new brain cells.[16]

Something similar also appears to apply to consciousness. It seem that our consciousness sacrifices some of its unbounded awareness each time we perceive the world through boundaries, which it does this to protect us from the rigors of daily life. The problem is that some of the things we do to protect ourselves end up distracting us, such as obtaining more and more possessions for an easier life. We need to transcend beyond the daily routine to keep expanding our consciousness. In short, we have to be careful of being limited to only what we know and experience. The *Kabbalah* echoes this when it states that spiritual and psychological wholeness comes from experiencing different dimensions of consciousness.

More scientifically, recall how Robert Becker and other scientists where able to reduce the level of consciousness of animals by applying opposite electric waves that cancelled out those of consciousness. If we have such waves in ourselves that start to cancel out others, we then reduce our own level of consciousness! Conversely, we can cohere and entrain waves together to increase consciousness.

As we know, nerve circuits and consciousness are important to the elevated elements of our mind and if they are growing in the right manner, so should our consciousness; we should more regularly become aware of our soul and spirituality as well as increase the likelihood of spiritual experiences. This is key to increasing Soul Power.

This is a little like letting your soul go to the gym to burn off unwanted fat. Many people exercise their bodies, but few exercise their souls beyond going to church.

We also need to tune in more with our unconscious, or at least subconscious and the perceptions that don't quite make it into consciousness. Some people call this listening to our intuition. This can provide a wider framework in which to make conscious decisions. For example, how often have you been able to tell that a person is negative or angry without even speaking to them?

Besides noting that unconscious thoughts precede conscious ones, Benjamin Libet found the mind approves or vetos unconscious impulses when they are made conscious. So while you may not be able to control what your unconscious is working on, you can control whether or not to act on it when it becomes conscious.

This choice to act or not is what provides us with self control and some element of free will, and is analogous to the moral compass ascribed by some theologians, and what some people attribute to the functioning of the soul.

The key point here is that we can consciously influence our perceptions as well as our responses to them. Remember how when we each read the word "yes", we each produce identical brain waves even though we might interpret this and other words slightly differently.

For example, in one study two groups of rats with cancerous tumors were given equal number of electric shocks. The only difference to the two groups was that one cage had a switch which the rats could push—and did each time a shock was administered. Even though pushing the switch did not stop the electric shock, the act of pushing it seemed to help this group, which had lower incidence of cancer and lower mortality rate compared to the other group which just had to accept the shocks.

"We ultimately have to shift our perceptions," says Rollin McCraty of the Institute of HeartMath. "The same events are going to happen. We're not going to stop the traffic jams. But we can modify how we're perceiving them."

He suggests one way to do this is to harness the heart. "The heart has a type of intelligence that's not cognitive, but more on the feeling level, more intuitive. It's part of why we feel our emotions in the heart. That's been recognized in folklore for centuries, but not in science until the modern recognition that the heart is one of the key points in our emotional system."[17]

He adds, "Intervening at the emotional level is often the most efficient way to initiate change in mental patterns and processes. Positive emotions not only feel better subjectively, but tend to increase synchronization of the body's systems, thereby enhancing energy and enabling us to function with greater efficiency and effectiveness. Individuals can maintain extended periods of physiological coherence through actively self-generating positive emotions."[18] To this end, the institute has developed a range of simple and effective programs that can alter the waves of your heart and brain. Their Freeze-Frame, Cut-Thru and other techniques are a very useful way of influencing heart and brain waves. Basically these technqiues use heart waves to prevent the brain from signaling the autonomic nervous system to shift into stress, panic or fear modes. Another key achievement is that they bring the power of emotions into the process to aid, cohere and strengthen change. (The institute's website www.heartmath.com provides details on these programs.)

By checking our breathing and pulse we can determine to some degree how the electric waves are resonating in our heads and hearts and how our energy is flowing. Being consciouslly aware that your sympathetic and parasympathetic nervous systems are sending different electric signals to reconcile stimuli and decide how to react to situations is a big step in not only controlling how we perceive things, but also in smoothing and cohering our brain and body waves, thereby reducing stress and its affects.

Another way to help smooth these waves is to recall your last major spiritual experience. This should re-ignite feelings of peace and calm that can help harmonize the waves. Similarly, thoughts and feelings of love (and even lust) can smooth and cohere waves.

McCraty found that during feelings of love, ECGs displayed harmonious sine wave patterns. "Many different organs and systems contribute to the patterns that ultimately determine our emotional experience. However, research has illuminated that the heart plays a particularly important role. The heart is the most powerful generator of rhythmic information patterns in the human body. Our data indicate that when heart rhythms patterns are coherent, the neural information sent to the brain facilitates cortical function. This effect is often experienced as heightened mental clarity, improved decision-making and increased creativity. Additionally, coherent input from the heart tends to facilitate the experience of positive feeling states. This may explain why most people associate love and other positive feelings with the heart," he says. "The heart is intimately involved in the generation of psychophysical coherence."

McCraty adds, "Numerous experiments have now demonstrated that the messages the heart sends the brain affect our perceptions, mental processes, feeling states and performance in profound ways. Our research suggests that the heart communicates information relative to emotional state (as reflected by patterns in heart rate variability) to the cardiac center of the brain stem (medulla), which in turn feeds into the intralaminar nuclei of the thalamus and the amygdala. These areas are directly connected to the base of the frontal lobes, which are critical for decision-making and the integration of reason and feeling. The intralaminar nuclei send signals to the rest of the cortex to help synchronize cortical activity, thus providing a pathway and mechanism to explain how the heart's rhythms can alter brainwave patterns and thereby modify brain function. Our data indicate that when heart rhythm patterns are coherent, the neural information sent to the brain facilitates cortical function. This effect is often experienced as heightened mental clarity, improved decision-making and increased creativity. Additionally, coherent input from the heart tends to facilitate the experience of positive feeling states. This may explain why most people associate love and other positive feelings with the heart and why many people actually "feel" or "sense" these emotions in the area of the heart."[19]

Another practical psychological technique to help overcome negative emotions and energy and improve your spirituality includes improving acceptance. This includes acceptance of the world around you, acceptance of yourself and that you are the one who determines your future, (while also being aware of the concerns of others).

Flexibility is another key element. With the world changing and energy continually fluctuating around you, flexibility enables you to bend with change rather than snap.

Other techniques to help change how you peceive things include practicing alternative or new behaviors, thoughts and feelings. Developing emotional intelligence may be more important than intellectual intelligence or IQ as to how you succeed in life, according to some people.[20]

Another useful tool is to take time out from perceiving at all. Moments of no sensation, or selflessness, provide a way for some people to get in touch with and grow their spirituality. Too often we seek something, rather than just be part of what already is. As Einstein noted, there are times when "one feels free from one's own identification with human limitation".

Conversely, many scriptures say we are always connected with the divine and know what is the right thing to do; we just forget how to connect and at other times we don't feel connected. Ancient scriptures to new age practitioners suggest we solve our problems by seeking the elements of God within ourselves. Some people say the right electromagnetic coherence can provide a connection to the universe. In contrast, jagged and conflicting waves make it harder to sense or determine what is the right thing to do.

Interestingly, psychologists suggest one reason why people feel isolated, weak and insecure is that they cut themselves off from their own energy, energy of others and the universe. To regain energy, they have to manipulate other people or dominate others to give them attention and thus energy and make themselve feel more powerful. This competition for human energy has even been described as the cause of all conflict between people.[21] Similarly, when we are playing a role or following a role model we are not connected to our own energy flows, but rather channeling flows to be that of someone or something else. While this can be useful to get us started in the spiritual direction we want to go and develop attributes we see in others we still need to connect to our own energy.

Another beneficial approach is to develop relationships that make such changes possible. Grouping with like people, such as positive spiritual people, can be beneficial. This is one element of how Alcoholics Anonymous and other organisations function. Choose a buddy or mentor with whom you feel comfortable discussing spiritual matters.

A coherent flow of energy among people can be beneficial for all involved. This can range from a whole community to a couple. For example, many of us know how powerful love between two people can be. This appears to be due to the coherence of one person's positive brain and body waves becoming entrained with and strengthening those of their loved one, as outlined in Part 1, and strengthened by an elaborate chemical process. Love, in its broadest sense pulls things together and is creative, note psychologists. Jung described love as "something superior to the individual, a unified and undivided whole".[22] Some people risk everything for the powerful energy of love. The issue of love also raises an intriguing side question: what is greatest, love or spirituality? You only have love for a lifetime, but you can have spirituality for eternity—or maybe we need to develop eternal love, a loving spirit that is the epitome of spirituality.

Right now, your own energy is worth something to those around and the environment around you. For example, when you leave your job you may be replaced in a few days, but you can never be replaced in the hearts and minds of your family and friends, (no matter how bad the relationships).

Relating with people is not always easy. Dealing with people is the hardest thing you will ever do, suggests the first person to climb Mt. Everest, Sir Edmund Hilary. This was harder for him than climbing the world's tallest mountain, he once told me. When dealing with people, take note of what they do more so than what they say.

Also, step back and look at the choices you are making, what you are doing and who you are relating to. Become consciously aware of why you are making those choices. Self-sabotaging beliefs and actions often creep into our lives, for a range of reasons, creating unwanted incoherence. Consider "spouse assassination" or where couples say negative things to each. While negative comments may initially be said jokingly, they can still impact the other person and eventually lead to self-questioning, doubt and incoherent brain and heart waves. In contrast, positive comments lead to more coherent and stronger waves. Similarly, if you think of yourself in negative terms, you are likely to influence your self-perceptions and undermine yourself and your spirituality. Learn to love thyself, as the scriptures suggest, and do not feel guilty if you are happy.

Remember, the universe does not distinguish good and bad energy, only we do. Similarly, there is no guilt in energy. There are only incoherent and jagged wave patterns that disrupt our bodies and minds.

The negative tags we apply to energy are much like the emotional tags we sometimes apply to our perceptions, learning and memories formed during LTP in order to better understand them. We are responsible for how we apply these tags. As Jung noted: "The individual who wishes to have an answer to the problem of evil, as it is posed today, has need, first and foremost of self-knowledge, that is, the utmost possible knowledge of his own wholeness. He must know relentlessly how much good he can do, and what crimes he is capable of, and must beware of regarding the one as real and the other as illusion. Both are elements within his nature, and both are bound to come to light in him, should he wish—as he ought—to live without self-deception or self-illusion. However, most people are hopelessly ill equipped for living on this level."[23]

Jagged electric waves can be made further chaotic by other negative perceptions and emotions. In this instance, the process results in a downward spiral, with waves being amplified more chaotically and stress being placed on the nervous system and body made worse.

We can turn "negative" energy or wave patterns around and into positive flows for ourselves, using wave principles, such as momentum, but it is up to us. I remember one motivational guru telling me the most important thing to look for in hiring people was their attitude. "You can train people to do almost anything, but you can't train their attitude." No one else can change your attitude and underlying spirit, it is up to you. Your attitude to life, friends, work and everything you do everyday is up to you.

This brings us to another important principle of waves to consider, acceleration. When waves accelerate they can also bump into waves in front of them, creating stronger ones. It is known that seals and sea lions can sense accelerating waves, such as those created by fish, reports German researcher Guido Dehnhardt.[24]

While the impact of accelerating electromagnetic waves in nerves and the body is little understood scientifically as yet, there would appear to be some influence. For example, increasingly thinking and feeling positive, happy and/or in love seems to build upon each other to create a powerful affect. While further research is required in this respect, it can't hurt trying to harness this principal of wave motion by combining electric waves, and accelerating them via emotions, to strengthen our electromagnetic wave network.

Whatever the case, by taking this Soul Power approach of taking greater control to integrate the different electromagnetic waves already flowing in ourselves, we take greater control of our lives and ultimately our spirituality. By being aware of, and understanding how the electromagnetic circuits of consciousness work we realize that we can influence those circuits.

This is analogous to being aware of your little toe. Most of the time you are not aware of it, but when you focus on it you can sense it. It is similar with your soul and spirituality; focusing on it will help you to be aware of it—and knowing now how it operates allows you to take greater control of it. For instance, recall how stochastic resonance enjoins electric waves with other random background waves to enable previously unrecognized sensations and perceptions to become conscious. Some people suggest music or waves created by consuming alcohol or other substances is enough to provide the background waves and invoke such stochastic resonance.

Also recall the research by Schwartz and others which demonstrated how with focus you can reshape your brain circuits—which suggests you can reshape your soul and spirituality.

There is increasing evidence of the brain's ability to remake itself throughout life, and this is not only in response to external stimuli, but also in response to our own mental effort, says Schwartz and Begley. "We are seeing, in short, the brain's potential to correct its own flaws and enhance its own capabilities," they report. "Physical changes in the brain depend for their creation on a mental state in the mind—the state called attention. Paying attention matters. It matters not only for the size of the brain's representation of this or that part of the body's surface, of this or that muscle. It matters for the dynamic structure of the very circuits of the brain and for the brain's ability to remake itself."[25] This suggests that we can each pull ourselves, our spirit, up by our own bootstraps.

Schwartz outlines how the quantum zeno effect of quantum physics could be involved in changing the circuitry of the brain when the mind focuses attention.

Under this effect, regular observation 'resets' probabilities (of uncertainties), so that as the number of observations increases the probability of a change in the quantum state decreases. This is analogous to saying that the watched kettle never boils.[26] In terms of the mind, this effect suggests that focusing attention keeps the circuits flowing—akin to the waves building upon electromagnetic waves in long-term potentiation.

Schwartz and Begley also propose the resulting behavior of the mind and ensuing actions of the human system can be influenced by what sort of attention is focused.[27] If you are afraid, for example, you are likely to unconsciously change your habits to avoid those disconcerting jagged waves that fear brings. However, you might not replace them with coherent wave patterns. You may just try to reduce the jagged waves to a less chaotic pattern and learn to live with that, even though it remains stressful to your system. So while the human system naturally seeks to reduce chaotic waves, you may have to consciously focus on ways to generate coherent waves that benefit it.[28]

This extends other research that shows fear of loss is greater than desire for gain—or that reduction of jagged wave patterns takes precedence ahead of generating coherent flows of electromagnetic wave. In turn, this suggests we have to not only practice thinking positively to become more positive, but also to be able to be receptive to experiencing things spiritual.

This is echoed in many religious scriptures, spiritual and new age texts; such as repeating various prayers, chants, the benefits of repetition of honest work and many more actions that focus attention in the patterns of electric waves inside us.

One of the best ways to become more aware of how your own waves are flowing is to develop new ways of seeing, such as by developing the ability to objectively look at ourselves from outside, as if through the eyes of an inpartial bystander. This can help provide a better understanding of how our electric waves and circuits are operating—and why the waves are the shape they are.

Then, we can take responsibility for "reprogramming" electromagnetic circuits beyond the simple stimulus-response reaction that dictate most of our everyday actions. As we reprogram ourselves, we can alter perceptions, separate stimulus from response, use the veto of consciousness and provide ourselves with more free will. We can remove the limitations and boundaries that are there. We can become more aware of ourselves and the world around us—and who we can become.

This leads us to the higher understanding that we are ultimately responsible to a great extent for ourselves and how we see the world. Or in other terms, the waves in electric circuits build upon each other to provide a greater self-awareness, self-realization or state of consciousness—Soul Power.

Taking the laws of energy, electromagnetism and wave principles into account in our daily lives can make a big difference, a Soul Power difference. Over time, you should experience a greater sense of calmness and it should gradually become a greater part of your day. Your reactions to difficulties should become less pronounced, less angry, and less prolonged. You should be able to bounce back from difficulties more readily as you realize they are less important to you and your life compared to your spirituality. This is not always easy as the modern world bombards our senses every minute, influences our perceptions and competes for our mental and spiritual attention. None-the-less, spiritual development today it is increasingly about what a person does and who they are.

Improving your Soul Power is generally not one big thing that you can do or change instantly. It takes a lot of small things, small things that add up to a great big whole. Don't be disillusioned that you have to do a lot of small things; in contrast, it is much easier to enact small changes than big ones. Weave the fragmented strands of yourself into a spiritual tapestry.

If you suffer a setback, learn from it. Choose a small, non-threatening task to prove yourself to yourself. Once you have succeeded in several small tasks, you will be ready to take on bigger challenges. Soon you will see that you are capable of doing whatever you put your mind and spirit towards. Though, remember, the bigger the challenges the more work you have to put in. For example, you can't climb a mountain without learning the skills of rock climbing and mountaineering. It takes a lot of smaller hills to summit before you can tackle Mount Everest.

Don't get discouraged. Embarking on the journey or pilgrimage itself provides enough purpose for some people, while for others, nothing short of unity with God or Allah is enough. This is despite Godel's incompleteness theorem suggesting that to seek self-knowledge is to embark upon a journey which will always seem to be incomplete.

This raises the question whether our souls continue once our nerve circuits have died?

The *Bhagavada Gita* suggests, "Our bodies are known to end, but the embodied self is enduring, indestructible, and immeasurable....as a man discards worn-out clothes to put on new and different ones, so the embodied self discards its worn-out bodies to take on other new ones. Weapons do not cut it, fire does not burn it, waters do not wet it, wind does not wither it."

Taoists believe humans lost the ability to achieve immortality due to having lost our state of harmony with nature, with the universe. Taoism and yogic tradition, state that at death, the soul emerges from the bregma, the junction of the sagittal and coronal sutures at the top of the skull, to merge with the world soul. This is called the Marrow Way by Chinese Taoists and the Way to Brahman by Indian Yogis.

Also, there are claims that the Bible alluded to reincarnation, but that this was deleted several centuries later by a pope who thought it would undermine the Catholic church, despite the obvious resurrection, or reincarnation, of Jesus.

Some scientists have also addressed this question. For example, Jammer reports that Einstein wrote to the family of a departed lifelong friend and suggested "for us faithful physicists, the separation between past, present, and future has only meaning of an illusion, though a persistent one".[29] This was just before Einstein's own death in 1955. Jammer adds, "Theologians and philosophers do not seem to know that the special theory of relativity itself, by means of its space-time geometrical diagrams, offers perhaps the best representation of the eternity-time relation".

Physicist Paul Davies notes, "If the mind is basically 'organized information" then the medium of expression of that information could be anything at all; it need not be a particular brain or indeed any brain. Rather than 'ghosts in machines', we are more like 'messages in circuitry' and the message itself transcends the means of it expression." He suggests, "This conclusion leaves open the question of whether the 'program' [our souls] is re-run in another body at a later date (reincarnation), or in a system in which we do not perceive as part of the physical universe (in Heaven?), or whether it is merely 'stored' in some sense (limbo)."[30]

Dutch scientist Pim van Lommel says his research into near death experiences indicates consciousness remains during a flat EEG. Memories of these experiences appear to come from the time when the brain is inactive, notes the cardiologist. In this intriguing study, medical equipment confirmed there was no pulse or apparent brain activity in the subjects and accordingly the newly dead should not have been able to perceive anything. However, almost one in five reported post—or near death experiences that were very structured. While the phenomenon defies medical explanation it does not defy that of physics, which says energy cannot be destroyed. Lommel suggests consciousness is independent of the white and gray matter of the brain. "Compare it with a TV program. If you open the TV set you will not find the program. The TV set is a receiver. When you turn off your TV set the program is still there, but you can't see it. When you put off your brain, your consciousness is there but you can't feel it in your body."

Researchers investigating consciousness "should not look in the cells and molecules alone," he suggests.[31]

However, other scientists believe near death experiences are mental fabrications like those of dreaming; in this case the mind tries to make sense of random neural firings as the brain dies, and in some cases switches back on and comes back to life. This is another story, requiring further research.

Whatever the case, we are more than who we think we are. As LeDoux suggests, "As we begin to understand ourselves in neural, especially synaptic terms, we do not have to sacrifice the other ways of understanding existence. A neural understanding of human nature broadens rather than constricts our sense of who we are."[32]

This new understanding of the electromagnetic basis of the soul and spirituality opens a whole new realm to investigate, learn and benefit from. But it is up to us whether we recognize this as just one note or part of a greater symphony. Understanding how the energy of a vibrating string of a guitar creates a sound that strikes our ears and is processed electronically in our brains does not detract from the joy of a pleasant piece of music, rather it can add another dimension to our appreciation—as can Soul Power.

Similarly, consider how the three-pounds of your brain contains as many neurons as the universe has stars; billions.

Consider the light waves of your image radiating throughout space, representing you in the universe, in a sense for eternity.

Then consider just what does stochastic resonance and spiritual experiences allow us to sense?

Some people say it is God. Some say it is nothing. Physics says it is something.

As science and spiritualists continue to investigate the soul, recall how you are the director of your own spirituality. Make each moment count. Become rather than just be.

You have the choice and power, the Soul Power, to alter brain and body waves, nerve circuits and perceptions, to transcend paradigms, to choose responses and ultimately shape your soul and spirituality. Choose well.

FOOTNOTES & REFERENCES

Introduction

1—As the presentation of subjective and controversial material is easily subject to personal bias, use has been made of quotations from the sources to avoid, as far as possible, any misinterpretations.

Chapter 1

1—Robert Broderick et al, *Catholic Encycolpedia* (1990).

2—Timothy Freke et al, *Encyclopedia of Spirituality* (2000).

3—*Encyclopedia Britannica online* (2002).

4—Where the most research has been undertaken in this respect.

5—Larry Culliford, in *British Medical Journal* 12/2002.

6—Max Jammer, *Einstein and Religion* (1999).

7—Joseph Campbell, in "The Power of Myth" series on *PBS TV* 1986.

8—Carl Jung, *Memories, Dreams, Reflections* (1983). Edited by Aniela Jaffe, translated by Richard & Clara Winston, copyright 1961, 1962, 1963 and renewed 1989, 1990, 1991 by Random House, Inc. Used by permission of Pantheon Books, a division of Random House, Inc,

9—Matthew Alper, *The God Part of The Brain* (2001).

10—Andrew Newberg, *Why God Won't Go Away: Brain Science and the Biology of Belief* (2000).

11—Global Consciousness Project, at noosphere.princeton.edu.

12—McKanna personal survey.

13—Daniel Dennett, *Consciousness Explained* 1991 and *Freedom Evolves* (2003).

14—Michael Shermer & Frank Sulloway, in *Humanist* 11/1999

Chapter 2

1—Tortora & Grabowski *Principles of Anatomy & Physiology,* (2000).

2—Harold Saxton Burr, *Blueprint for Immortality* (1972).

3—Candace Pert, media interview at www.primalpage.com/pert.htm 1995.

4—Ibid

5—Jesse Roth, National Institute of Health in the 1980s, in *Smithsonian* 11/1998.

6—Candace Pert, *Molecules of Emotion* (1997).

7—David Waters, in *The Commercial Appeal* 1/2001.

8—As discovered by Jesse Roth, at the National Institute of Health in the 1980s, in *Smithsonian* 11/1998.

Chapter 3

1—Tortora & Grabowski, *Principles of Anatomy & Physiology,* 2000.

2—It is uncertain if this is a way that mass is created from energy. It also appears that this force has time element integrated to it that is not yet well understood.

3—John Taylor, *Hidden Unity in Nature's Laws* (2001) and Issac Asimov, *Atom* (1992).

4—John Taylor, *Hidden Unity in Nature's Laws* (2001).

5—Rod MacKinnon, in *Nature* 2001

6—Owen Hamill et al, in *American Scientist* 1/1995; and Tortora & Grabowski *Principles of Anatomy & Physiology* 2000.

7—The resting membrane potential of nerve cells has been found to range from—40 to—90 milliVolts, with a typical charge of—70mV. The minus sign indicates that the inside is the negative relative to the outside. The range for polarized body cells is +5mV to—100mV, depending upon the cell.

Chapter 4

1—Tortora and Grabowski, *Principals of Anatomy and Physiology* (2000).

2—Candace Pert, *Molecules of Emotion*; Joseph LeDoux, *Synaptic Self* (2002); and *Science* 2/1999.

3—Charles Gray and David McCormick, in *Science* 10/1996.

4—Michael Rudolph et al, in *Journal of Computational Neurosceince* 11/2001; Roger Traub, in *Science* 10/1996.

5—Joseph LeDoux, *Synaptic Self* 2002. Followers of biologist Robert Becker suggest the shorter route involves continuous current signaling, while the higher road and involves digital signaling of neurons—see later chapters.

6—Some scientists, such as Nobel Laureate Francis Crick, believe it takes a frequency of 10 Hz to initiate LTP in human nerve cells, while others note the chattering pyramidal cells resonate at 40Hz. Further research is obviously required to clarify this range.

7—Joseph LeDoux, *Synaptic Self* (2002) and many others.

8—James Weaver and Dean Astumian, in *Science* 1/1990 as well as research by Kenneth J. McLeod, State University of New York, Stony Brook.

9—Arthur Pilla, in *Bioelectrochemistry and Bioenergetics* 2/1998.

10—Tortora & Grabowski *Principles of Anatomy & Physiology* (2000).

11—Laurie Silva et al, in *Science* 1/1991.

12—James Ryaby et al of Orthologic at www.orthologic.com.

13—John Clay & Alvin Shrier, in *Biological Bulletin* October 2001.

14—Joseph LeDoux, *Synaptic Self* (2002).

15—James Oschman, *Energy Healing: The Scientific Basis* (2000), courtesy of Churchill Livingstone.

16—Gyorgi Buzsaki, of Rutgers University, in *Proceedings of the National Academy of Sciences* 3/2003 as reported in *Discovery News.*

17—Terrence Sejnowski, Salk Institute of Computational Neurobiology Labatory.

18—Joseph LeDoux, *Synaptic Self* (2002).

19—James Brewer and Anthony Wagne, in *Humanist* 1/2002.

20—Vadim Bolshakov, William Carlezon and Eric Kandel, in *Neuron* April 2002.

21—As reported in *Western Daily Press* 11/2002.

22—Joseph LeDoux, *Synaptic Self* (2002).

23—Harold Saxton Burr, *Blueprint for Immortality* (1972).

24—Andrew Newberg, *Why God Won't Go Away* (2002).

25—Joshua Greene, in *Humanist* 1/2002.

26—Wilder Penfield et al, *Mystery of the Mind: A Critical Study of Consciousness and the Human Brain* (1975).

27—Jose Delgado, *Physical Control of the Mind* 1969 and Delgado in *Free Inquiry* 11/1995.

28—Peter Hauri, in *Newsday* 5/1991.

Chapter 5

1—Tortora and Grabowski, *Principals of Anatomy and Physiology* (2000).

2—J. Andrew Armour, *Neurocardiology* (1994); and Institute of HeartMath.

3—Michael Gershon, *The Second Brain* (1999).

4—Tortora and Grabowski, *Principals of Anatomy and Physiology* (2000).

5—The initial sound waves are believed to be comprised of vibrating atomic lattices called phonons.

6—Odorous substances also show a displacement between incident and reflected light spectra known as the Raman shift, which demonstrates a shift in energy.

7—Sensations such as pressure, heat, tickling and more can be produced by varying the chronaxie, an electrical pulse.

Chapter 6

1—Stanislas Deheane et al, in *Sunday Times* 10/2002

2—Antonio Damasio, *The Feeling of What Happens: Body and Emotion in the Making of Consciousness* (1999).

3—Michael Gershon, of Columbia University College of Physicians and Surgeons, *The Second Brain* (1999).

4—Marcus Raichle of Washington University, in *US News & World Report* 11/2001.

5—Francis Crick, in *UK Sunday Times* 11/1998.

6—Fransisco Varela of Salpetriere Hospital in Paris, in *Newsday* 2/1999.

7—John Hart et al, of the University of Arkansas for Medical Sciences in university news release 5/2002.

8—Joseph LeDoux, *Synaptic Self* (2002).

9—Susan Greenfield, neuroscientists at Oxford University, *Journey to the Centers of the Mind* (1995) and *The Private Life of the Brain* (2001), in *Financial Times*, 6/2000.

10—Roger Penrose at al, *Large, the Small and the Human Mind* (2000).

11—John Taylor, *Hidden Unity in Nature's Laws* (2001).

12—James Glanz, in *Science* 9/1997 and James Collins et al, in *Nature* 10/1996.

13—Joachim Gartzke, in *American Journal of Physiology* 11/2002.

14—Frank Moss, University of Missouri, in *The Economist* 9/1998.

15—Ichiro Hidaka et al, in *Physical Review Letters* 10/2000; and Bruce Gluckman et al, in *Physical Review Letters* 11/1996.

16—Benjamin Libet of the University of California, in *National Post* 1/2003 and other media.

17—Jose Delgado, *Physical Control of the Mind* (1969).

18—Joseph LeDoux, *Synaptic Self* (2002).

Chapter 7

1—Robert Becker, *The Body Electric* (1985).

2—Ibid.

3—Understanding one does not provide the systemic knowledge that comes from understanding both and how they relate.

4—Robert Becker, *Cross Currents* (1990). Copyright 1990 by Robert O. Becker. Used by permission of Jeremy P. Tarcher, an imprint of Penguin Group (USA) Inc.

5—Benjamin Libet, of the University of California, in *National Post* 1/2003 and various other media.

6—Robert Becker, *Cross Currents* (1990).

7—Ibid.

8—This was in the 1940s. In Robert Becker, *The Body Electric* (1985).

9—Robert Becker, *The Body Electric* (1985).

10—Catherine Verfaille, University of Minnesota, in *The Guardian* 6/2002.

11—Betty F. Siskin et al, in *Bioelectromagnetics Society* 6/1998.

12—Robert Becker, *Cross Currents* (1990).

13—Jose Delgado, *Physical Control of the Mind* (1969).

14—Robert Becker *The Body Electric* (1985).

15—Shandala & Vinogradov in, www.emraa.org.au/rf/microwaving.htm.

16—Valerie Hunt, *Infinite Mind: Science of the Human Vibrations of Consciousness* (1996), courtesy of Dr. Hunt.

17—Robert Becker, *The Body Electric* (1985).

18—Robert Becker, *Cross Currents (*1990).

19—A. M. Sinyukhin of Lomonosov State University, in Harold Saxton Burr, *Blueprint for Immortality* (1972) courtesy of the C.W. Daniel Company Ltd.

20—Harold Saxton Burr, *Blueprint for Immortality* (1972).

21—Ibid.

22—Robert Becker, *The Body Electric* (1985).

23—Valerie Hunt, *Infinite Mind* (1996).

24—Robert Becker, *Cross Currents* (1990).

Chapter 8

1—Wilder Penfield et al, *Mystery of the Mind: A Critical Study of Consciousness and the Human Brain* (1975).

2—Michael Persinger, on *ABC News Nightline* 1/2002.

3—Michael Persinger, in *Free Inquiry* 1/1996.

4—Atheists are people who believe there is no God, while agnostics tend to be people who believe it is impossible to know whether there is a God.

5—Michael Persinger, *Neurotheology* (2002).

6—Michael Persinger, in *Free Inquiry* 1/1996.

7—Michael Persinger et al, *Neuropsychological Bases of God Beliefs* (2002).

8—Ibid.

9—VS Ramachandran of the University of California, in *BBC News.com* 3/2003.

10—Olaf Blanke of Geneva University Hospital, in *Nature*, 2002.

11—Peter Brugger of the University Hospital of Zurich, in *Chicago Tribune* 7/2000.

12—John O'Keefe et al, *The Hippocampus as a Cognitive Map* (1978).

13—Michael Persinger at al, *Neuropsychological Bases of God Beliefs* (2002).

14—Michael Persinger et al, *Perceptual & Motor Skills*, 6/2001.

15—Michael Persinger et al, in *Perceptual & Motor Skills*, 2/2001.

16—Robert Becker, *The Body Electric* (1985).

17—Michael Persinger in *Neurotheology* (2002).

18—Robert Becker, *Cross Currents* (1990).

19—Richard Frankel, California Polytechnic University at San Luis Obispo, in *Science* 5/1991.

20—Mark George of the University of South Carolina, in *Newsweek* 6/2002.

Chapter 9

1—Houston Veterans Medical Center 1994, in the *Post Standard Syracuse* 9/2000.

2—Howard Brody, *The Placebo Response* (2001).

3—Daniel Goleman, in *The San Francisco Chronicle* 7/1988.

4—University of California—Los Angeles news release.

5—Maryanne Garry, Victoria University in New Zealand, in *BBC News online* 1/2003.

Chapter 10

1—*Bible* 1 John 4:24.

2—Gary Zukav, *Seat of the Soul* (1999) & *The Heart of the Soul* (2002).

3—From Institute of HeartMath's website—www.HeartMath.com.

4—Ibid.

5—Rollin McCraty, *The HeartMath Report* (2001).

6—Joseph LeDoux, *Synaptic Self* (2002).

7—From Institute of HeartMath's website—www.HeartMath.com.

8—Richard Davidson, *Neurobiology of Learning and Memory* 1/2001.

9—Joseph LeDoux in Robert Conlan et al, *States of Mind* (1999).

10—Joseph LeDoux, *Synaptic Self* (2002).

11—David Diamond, Bruce McEwen and others.

12—Joseph LeDoux, *Synaptic Self* (2002).

13—Press release from the Netherlands Organization for Scientific Research 1/2002.

14—Rhawn Joseph, *Transmitter to God* (2001).

15—Joseph LeDoux interview at www.thirdedge.org.

16—Rhawn Joseph, *Neurotheology* (2002).

17—Ibid.

Chapter 11

1—Thomas Moore, *Care of the Soul* (2002).

2—Carl Jung, *Memories, Dreams, Reflections* (1963).

Chapter 12

1—Interestingly, a photon that travels at the speed of light has no mass.

2—John Taylor, *Hidden Unity in Nature's Laws* (2001), and Bernard Haisch, Alfonso Rueda and H.E. Puthoff, in *Dallas Morning News* 4/1994.

3—Janusz Slawinski, in *Brain-Mind Bulletin* 10#9 1985.

4—Valerie Hunt, *Infinite Mind* (1996).

5—Frank Schubert, in *Encounter* (1984).

6—John Taylor, *Hidden Unity in Nature's Laws* (2001).

Chapter 13

1—These waves also travel through more than one plane, that is not just one direction, but rather more like a pulse.

2—The energy conveyed by an electromagnetic wave is proporational to the frequency of the wave. The wavelength and frequency of the wave are connected via the speed of light. www.physics.bu.edu.

3—Discovered by Thomas Young in 1801.

4—Sharon Begley, *Newsweek* 6/1995.

5—Lene Vestergaard of Harvard University, in *Los Angeles Times* 1/2001

6—Is this what happens with homeopathy?

7—David Bohm, *The Undivided Universe* (1993).

8—Brian Greene, *The Elegant Universe* (1999).

9—Ibid.

10—Stephen Hawking, *A Brief History of Time* (1989).

11—Brian Greene, *The Elegant Universe* (1999).

12—More details on this fascinating topic can be found in Amir Aczel, *Entanglement* (2002).

13—Frank Moss, University of Missouri, in *The Economist* 9/1998.

14—Ichiro Hidaka et al, in *Physical Review Letters* 10/2000; James Glanz, in *Science* 9/1997; Bruce Gluckman, in *Physical Review Letters* 11/1996; and James Collins et al, in *Nature* 10/1996; among others.

15—Michael Rudoplh et al, in *Journal of Computational Neuroscience* 5/2001.

16—Ibid.

17—Jim Collins at Boston University and Louis Lipse of Boston's Hebrew Rehabilitation Center for the Aged, in *All Things Considered* 9/1998.

18—Ichiro Hidaka et al, in *Physical Review Letters* 10/2000.

19—Dean Astumian's, of the University of Chicago, "Brownian Ratchet" model of increasing electrical noise provides a method to understand how cells can detect and respond to weak external fields.

Chapter 14

1—Mantak & Maneewan Chia, *Awaken Healing Light of the Tao* (1993), courtesy of Mantak Chia and the Universal Tao Center.

2—Maki Takata, Toho University, 1938.

3—Paul Davies, in *New Scientist* 11/2001.

4—Dalai Lama, *A Simple Path* (1997).

5—Max Jammer, *Einstein and Religion* (1999), courtesy of Princeton University Press.

6—*Discover* 6/2002.

7—David Bohm, *Science, Order and Creativity* (1987).

Chapter 15

1—As per Dirac's field theory.

2—John Taylor, *Hidden Unity in Nature's Laws* (2001).

3—According to Einstein's theories, energy is the source of gravitation and of space-time curvature, in John Taylor's *Hidden Unity in Nature's Laws* (2001).

4—Harold Saxton Burr, *Blueprint for Immortality* (1972).

5—Bernard Haisch et al, in *The Sciences* 11/1994.

6—David Bohm, *The Undivided Universe (*1993).

7—Paul Davies, in *New Scientist* 11/2001.

8—David Bohm, *The Undivided Universe* (1993).

9—Amir Aczel, *Entanglement* (2002).

10—David Bohm in an interview with Renee Weber of Rutgers University. From Renee Weber, *Dialogues wuith Scientists and Sages: The Search for Unity* (1986),

11—Ibid.

12—Bruce Gluckman et al, in *Physical Review Letters* 11/1996.

13—David Bohm, *The Undivided Universe* (1993).

14—As a speculative aside as to how this might work in terms of quantum physics, consider a source of waves. There are waves spreading out in nearly infinite directions with as many probabilities of actions. As one or several waves interact with other waves they create a reaction and the probability becomes a reality. This interaction includes measurement and perception. Those that do not interact or are "bounced off" a field become quantum fuzz. This applies to just one spectra, one element of wave motion (be it frequency or other defining element). There are similar interactions for different elements of the wave spectrum. Some will apply to only one spectra, while others will cross many parts of the spectrum, such as how lead can stop may elements of wave radiation, by creating a field that spans parts of the spectrum. Also, it appears that the composition of matter also has an impact on waves, with some waves flowing through the ether or air much easier than through solid objects and vice versa. Waves of photons can obviously flow through the ether and air much easier than through the atoms of a solid object. There also appears to be a relationship between the number or intensity of waves and the impact they have on a field, sometimes one wave can exert an influence while it may take many waves to effect a wave interaction and change in the field. Also, in terms of our perception, consider another example whereby various waves of the source are sensed by our eyes, ears, hearing, taste, touch and maybe even pineal gland and other organs. These senses each report a slightly different form of wave either by position or wavelength or other measure. Our nervous system, including the brain, recompiles these waves using electric circuits to provide our image of that wave source. As we do this, we can add emotions (waves and chemicals) and wave memory to the electric waves going around the circuit. This is how we can develop and recall fears with associated objects and actions: we attach our fears (jagged waves) to the perception. There is no such energy with the original object being perceived, but because of our situation, our mental processes, we can and often do attach additional information to the sensing and recoding of simple physical waves. Under quantum physics, the act of perception creates an actual reality, which we in turn perceive as a reality specific, or as Einstein would say, relative to us. All this produces its own quantum effect in the waves of our electric circuits. As we understand the quantum interaction of waves and fields better, we will no doubt also begin to better understand how electromagnetic waves work in respect of ourselves and our spirituality. In the meantime, this theory attempts to make it easier for the lay person to understand how waves can influence us and how we can in turn influence them to our benefit.

15—Kenneth McLeod et al, in *Science* 6/1987.

16—James Weaver et al, in *Science* 1/1990.

17—Ibid.

18—Fatih Uckun, Wayne Hughes Institute, in *Science News* 2/1998; and in *Journal of Biological Chemistry.*

19—Herbert Frohlich, *Biological Coherence* (1988).

20—Betty F. Siskin et al, in *Bioelectromagnetics Society* 6/1998.

21—Valerie Hunt, *Infinite Mind* (1996).

22—Harold Saxton Burr, *Blueprint for Immortality* (1972).

23—Bill Lenth, in *Optical Memory News* 4/1994.

24—Karl Pribram, director of the Radford Brain Research Center, in *The Globe and Mail* 12/1996.

25—David Bohm, *Wholeness and the Implicate Order* (1996).

26—W. Pannenberg, *The Doctrine of Creation and Modern Science* (1988).

27—Harold Saxton Burr, *Blueprint for Immortality* (1972).

Chapter 16

1—Pitzer College California, in *Wall Street Journal* 6/2002.

2—Michael Meaney, Canada's McGill University, in *Montreal Gazette* 12/2002 and other media.

3—Paul Davies, *God and the New Physics,* (1983). Reprinted with the permission of Simon & Schuster. Copyright 1983 by Paul Davies.

4—Some sceptics say spiritual experiences involve nothing other than the sensing of our own mental manifestations, how do they explain the sensation of love? While love cannot be seen or described tangibly, it is a major part of most of our lives—a holistic experience with real electromagnetic,chemical, physical and other properties that can be described and which each have an impact.

5—Andrew Newberg & Eugene d'Aquili, *Why God Won't Go Away: Brain Science and the Biology of Belief* 2001.

6—Harold Saxton Burr, *Blueprint for Immortality* (1972).

Chapter 17

1—The Templeton prize.

2—The Vatican and the Center for Theology and Natural Sciences has sponsored several conferences on the relationship between the soul and the brain, including scientific aspects.

3—Albert Einstein, in Max Jammer, *Einstein and Religion* (1999).

4—Ralph Waldo Emerson went even further and said, "the religion that fears science insults God and commits suicide."

5—Max Jammer, *Einstein and Religion* (1999).

6—Candace Pert, *Molecules of Emotion* (1997).

7—John Polkinghorne, in *Hamilton Spectator* 10/2002.

8—*The Times* (UK) 6/2002.

9—Leon Lederman et al, *From Quarks To Cosmos* (1989).

Chapter 18

1—In this case, preparation and physics of metal fatigue.

2—Interestingly, Taoism states that chakras, contrary to new age perceptions, have no individual power and are points through which larger forces flow. To increase the power of one chakra requires more energy flowing throughout the body's entire chi system. There are some indications that the chakras are related to the enochrine system, in particular the major nerve and glandular centers of the body, according to author Richard Gerber.

3—Tortora & Grabowski, *Principles of Anatomy & Physiology* (2000).

4—Robert Becker, *Cross Currents* (1990).

5—As described in Part I.

6—Valerie Hunt, *Infinite Mind* (1996).

7—Tortora & Grabowski, *Principles of Anatomy & Physiology* (2000).

8—Russell Beckett, on his website www.uniquewater.com.au.

9—Russel Beckett, in *Sydney Morning Herald* 4/2002.

10—Jean Benveniste, *Digital Biology* (1998).

11—Rolland Conte et al, *Theory of High Dilutions and Expermental Aspects* (1996).

12—C.Z. Hong, in *Archives of Physical Medicine and Rehabilitation* (1982).

13—*Water Technology News* 6/1997.

14—Carlos Vallbona, Baylor College of Medicine, in *New York Times* (1997).

15—*Water Technology News* 6/1997.

16—Robert Becker, *The Body Electric* (1985).

17—Richard Frankel, California Polytechnic University at San Luis Obispo, in *Science* 5/1991.

18—Deepak Chopra *Quantum Healing* (1989).

19—Robert Becker, *Cross Currents* (1990).

20—Rollin McCraty et al, in *Proceedings of the Tenth International Montreux Congress on Stress, Montreux, Switzerland* 1999.

21—Candace Pert, preface to James Oschman's *Energy Medicine: The Scientific Basis* (2000).

22—James Oschman, *Energy Healing: The Scientific Basis* (2000).

23—J. Chien et all, in *American Journal of Chinese Medicine* 6/2001.

24—Charles Alexander, in *Journal of Personality and Social Psychology* 1980.

25—University of Michigan, in *Science News* 6/1976; and David Holmes, University of Kansas in *Skeptical Inquirer* 5/1995.

26—Andrew Newberg et al, in *Neurotheology* (2002).

27—James Oschman, *Energy Healing: The Scientific Basis* (2000).

28—Judith Lasater, in *Yoga Journal* 5/1999.

29—1993 British study reported in www.yogajournal.com.

30—R.M. Mattijs Cornelissen of Sri Aurobindo Ashram in Pondicherry, India, as reported in *Yoga Journal*.

31—Institute of HeartMath and others.

32—Joshua Rosenthal & Francisco Bezanilla, in *Biological Bulletin* October 2000.

33—Valerie Hunt, *Infinite Mind* (1996).

34—notably Thelma Moss.

35—Kirlian Photography Research website 2002.

36—Gary Schwartz & Linda Russek, University of Arizona *U-Wire* 6/2001; and various media reports.

37—Carl Jung, *Memories, Dreams, Reflections* (1963).

38—For example, recall the great pyramid razor sharpening fad. Research has since shown that this had nothing to with the shape of the pyramid, but rather than it helped to dry razorblades, which are blunted slightly by droplets of water.

Chapter 19

1—Plato, *Dialogues.*

2—Norman Cousin, *Anatomy of an Illness* (1979).

3—Deepak Chopra *Quantum Healing* (1989).

4—Robert Becker, *Cross Currents* (1990). Becker also notes that when chemotherapeutic drugs are administered during a patient's biocycle is a major determiner of their effect. Given at the appropriate time, they are more effective against cancer cells and produce fewer side effects than if given at the wrong time.

5—Valerie Hunt, *Infinite Mind* (1996)

6—S. Smith et al, in *Transactions of Biomedical Engineering* 1986; and A. Surowiec et al., in *Transactions of Biomedical Engineering* 1987.

7—Xin Yuling, in *China Daily* 11/1998.

8—Robert Becker, *Cross Currents* (1990).

9—U.S. National Institute of Health.

10—Carol Rausch Albright, in *Neurotheology* (2002).

11—Russek & Schwartz, in *LA Daily News* 3/1997.

12—Deepak Chopra *Quantum Healing* (1989).

13—David Spiegel, Lydia Temoshok, et al, in *American Health* 11/94.

14—Joseph LeDoux in Robert Conlan et al, in *States of Mind* (1999).

15—Candace Pert, *Molecules of Emotion* (1997).

16—Mantak & Maneewan Chia, *Awaken Healing Light of the Tao* (1993).

17—In an interesting aside, one study found a correlation between age of reproduction and life expectancy. An experiment by Michael Rose of University of California at Los Angeles found that the later in life that fruit flies reproduced, the longer they lived. By selectively breeding those that reproduced later in life, he developed flies that lived 130 days instead of the usual 40. In another approach, Cynthia Kenyon of the University of California at San Francisco doubled the life spans of nematode worms by removing their cells that produce eggs and sperm.

18—Kuo Mo-jo, India, c900–200 BC.

19—Valerie Hunt, *Infinite Mind* (1996).

20—Roy Bakay of Ruch Presbyterian-St. Luke's Medical Center in Chicago, in *Newsday* 6/2001.

21—We won't discuss electromagnetic pollution, here, as caused by the increasing amount of electromagnetism in the world we live in. While uncertain, it appears this has to be very low frequency and very localized (i.e. focused inside our head) to have an impact on our spirituality.

Chapter 20

1—Joseph Campbell said he believed a major issue for religion today was to awaken the heart. The task of a church is to provide people with an opportunity to identify with the God that is within each one of us, he suggested, says Wayne Holst in *National Catholic Reporter* 12/2001.

2—Thomas Moore, *The Soul's Religion* (2002).

3—Pope John Paul II, in *US News & World Report* 1/2000.

4—Albert Einstein, *What I Believe.*

5—Max Jammer, *Einstein and Religion* (1999).

6—Ibid.

7—Ibid

8—Paul Davies, *The Mind of God* (1992).

Chapter 21

1—Paul Davies, *The Mind of God* (1992).

2—Max Jammer, *Einstein & Religion* (1999).

3—Andrew Newberg, *Why God Won't Go Away* (2002).

4—Matthew Alper, in *Neurotheology* (2002).

5—Joseph Campbell, in *PBS TV interview* 1987.

6—Some people suggest this describes how we lost the contentedness of Adam and Eve in the Garden of Eden, gaining in its place Catholicism's 'original sin' as the consequence of tasting the fruit of knowledge. We yearn to recover that contentedness of ignorance without renouncing the world and knowledge. Daniel Matt in his introduction to the Kabbalah. As for how to do this, the Kabbalah says: "Though life branches out further and further, everything is joined to Ein Sof, included and abiding in it. Delve into this. Flashes of intuition will come and go, and you will discover a secret here. If you are deserving, you will understand the mystery of God on your own." This is analogous to descriptions of the string theory and the four fundamental forces of the universe, but just an analogy. Unfortunately, it does not provide any detail on how we might discover the secrets of these forces.

7—Mantak Chia, *Awaken Healing Light of the Tao.*

8—This would indicate that the ying-yang pattern of balancing positive and negative energy is not a model that should be considered for the soul today. There is no need to balance positive energy, as this is done naturally in today's modern world, with so much negative energy outside our bodies that we cannot keep up developing positive energy to protect ourselves from it.

9—Joseph Campbell, in *PBS TV interview* 1987.

10—Carl Jung, *Memories, Dreams, Reflections* (1963).

11—Ibid.

12—James Fowler, *Stages of Faith* (1995).

13—Valerie Hunt, *Infinite Mind* (1996).

14—Max Jammer, *Einstein & Religion* (1999).

15—Valerie Hunt, *Infinite Mind* (1996).

Chapter 22

1—Joseph LeDoux, *Synaptic Self* (2002).

2—To do so, stop and take that time out. Once you have cleared your mind, think about the three words that best sum you up. Then another three that you think describe your soul and spirituality. This is not always easy. If you have trouble, consider the principals that govern the way you act and think. Are they related to your spirit? Who do you think you are, is another good question as it leads to three others; who do you want to be, who don't you want to be, and where you currently fit in between these two? Similarly, when you walk do you walk fast, purposely or slowly amble? Do you always just go with the flow, where forces push you? Or do you try to take some control and move yourself towards where you want to go in life? Think about it, as the way you do things, such as walk, can reflect your approach to life and your Soul Power. Once you have determined three words that best, and truthfully, describe your soul and spirituality write them down at the top of a page. At the bottom, think about and write down three words you would like people to say about you and your spirit. Consider what those three words at the bottom really mean to you. Over the next few days, ask your family and friends what three words come to mind when they think of you. (Don't prompt them!) Be sure to ask them what they think about your soul or spirituality, rather than yourself. Would they say things such as "very spiritual," "honest," or "nasty," "vindictive," or something else? Often, people will just describe what they think of you, so it can be best to ask them for three words to describe you, then another three to describe your soul. Once you have these latter three words look them up in the dictionary to make sure you understand what they actually mean. Are you happy with these words? If you are, congratulations. More than 99 out of 100 people are not. Most people want to improve them—and their spirituality. Taking the words you have from other people, put the negative ones on the left and the positive ones on the right. These words provide a framework of what Soul Power you are currently projecting. This is one of the best ways to identify your Soul Power, as it involves other's telling you their perception of you and your spirituality, not just your own opinion. Couple these words with your own words that you already have at the top and the bottom of the page and you now have a rough map of where you are, and where you want to and need to go in life—and your Soul Power plan. In the middle of the page add the three principles that guide you though life. Answers to the earlier questions can also help provide greater clarity. This plan provides an idea of where you are, where you need to go, and a tool of how to get there. For example, acknowledge the negatives and work on changing them, while enhancing the positives. You can also use the principles and words as a guide to help you perceive and reconcile uncertainties and emotional events. In a moment, we will learn about some physical principles of energy that provide ways to achieve this. With this plan, the physics of energy and just a few minutes of focus on your spirituality a day you can influence things that we each do hundreds of times a day (but which most people do nothing about) to develop your Soul Power to make a difference. Unlike personal development fads that come and go, Soul Power is based on unchanging fundamentals of life. It provides a holistic way to take control of your life.

Chapter 23

1—Joseph LeDoux, *Synaptic Self* (2002).

2—Institute of HeartMath, at www.HeartMath.com.

3—Ed Diener, professor of psychology at the University of Illinois, and author *of Culture and Subjective Wellbeing* (2000).

4—David Myers, *The Pursuit of Happiness* (1993).

5—Martin Seligman, author of *Authentic Happiness,* in *USA Today* 12/2002.

6—Christopher Peterson of the University of Michigan, in *USA Today* 12/2002.

7—Dale Carnegie, *How to Win Friends and Influence People* (1982).

8—Richard Gerber, *Vibrational Medicine* (2001).

9—Richard Davidson of the University of Wisconsin, in *Newsday* 2/2003.

10—Richard Davidson, in *Los Angeles Times* 10/1996.

11—Media conference at the University of Wisconsin's Fluno Center 5/2001.

12—Joseph LeDoux, in *Sydney Morning Herald* 1/2003.

Chapter 24

1—From Institute of HeartMath's website at www.HeartMath.com

2—As outlined in Part I.

3—Tim Kasser of Knox College, *The High Price of Materialism* (2002).

4—Marilyn Albert of Harvard Medical School, in *Psychology Today* 11/1996.

5—Jeffrey Schwartz of the University of California Los Angeles and Wall Street Journal science writer Sharon Begley, *The Mind and The Brain: Neuroplasticity and the Power of Mental Force* (2002).

6—As cited earlier in Bob Beck, Richard Gerber, Andrew Newberg and Harold Saxton Burr.

7—Gallup Polls, in *Houston Chronicle* 12/1997.

8—Interestingly, studies at the University of Massachusetts Medical Center and University of Western Ontario both found that people who meditated regularly had higher levels of melatonin compared to those who took 5mg supplements. Recall the role of melatonin and the pineal gland as outlined in Part I. Reported in *Psychology Today* 11/1996.

9—Some affirmations with a spiritual element include:

- Change is inevitable, growth is optional
- Change how you see, not how you look

- Energy flows where focus goes
- Minds are like parachutes, they only function when open
- A goal without a plan is just a wish
- The best way to predict the future is to help create it
- Are you part of the problem, or part of the solution?
- Count your blessings not your troubles
- Yesterday is history, tomorrow is a mystery, today is a gift
- Believe in life before death
- Beautiful young people are accidents of nature, but beautiful old people are works of art
- One of the easiest, yet also the hardest, things to do is be nice to people
- Not all who wander are lost
- You will find adventure or adventure will find you.
- The truth is out there, don't get stuck here
- If someone betrays once, it is their fault; if they betray you twice, it's your fault
- When you lose, don't lose the lesson. Learn from your mistakes
- Judge your success by what you have to give up in order to achieve it
- Sometimes not getting what you want can be fortunate
- Ignorance is no excuse
- God is too big for one religion
- Anger is only one letter short of danger
- The greatest thing to fear is fear itself

- Learn from the mistakes of others, you can't live long enough to make them all yourself

- You use dental floss, so don't forget to use mental and spiritual floss

- Great minds discuss ideas, average minds discuss events, small minds discuss people

- Remember the three Rs—respect for yourself, respect for others and responsibility for your actions (spending time alone can help you learn to live with, and respect, yourself)

- Many people walk in and out of your life, but only true friends leave footprints in your heart

- Great love and great achievement involve great effort.

- A loving atmosphere in your home is the foundation for a love filled life. (In disagreements with loved ones, deal only with the current situation, don't bring up the past)

- Going to the garage does not make you a car, similarly going to church does not necessarily make you spiritual

- Your life is what your thoughts and you make it

Compiled from unattributed various sources, predominantly widely circulated anonymous emails.

10—Richard Gerber, *Vibrational Medicine* 2001 and others as noted earlier. Of the various touch therapies examined, Jin Shin Jyustu appears to be among the most grounded in science.

11—Sabrina Mesko, *Healing Mudras: Yoga for Your Hands* (2000).

12—For several years, I studied represenations of Buddhas in musuems around the world and noted the commonality, evolution and distortion of these positions. Some certainly seem to work, especially the "tranquility" pose of the serene Budda from Japan in New York's Metropolitan Museum of Art, while others have little affect on me. Is this just a placebo affect?

13—Carl Jung, *Memories, Dreams, Reflections* (1963).

14—Max Jammer, *Einstein and Religion* (1999).

15—Changes to the anterior cingulate were reported by Richard Davidson, in *Americal Journal of Psychiatry* 1/2003.

16—Fred Gage of the Salk Institute, in *Sydney Morning Herald* 2/2003.

18—Rollin McCraty, *The HeartMath Report* (2001).

19—Ibid.

20—Daniel Goleman, *Emotional Intelligence* (1997).

21—James Redfield, *The Celestine Prophecy* (1997).

22—Carl Jung, *Memories, Dreams, Reflections* (1963).

23—Ibid.

24—Guido Dehnhardt, of the University of Bonn Germany, in *The U.K Times* 7/1998.

25—Schwartz et al, *The Mind and The Brain: Neuroplasticity and the Power of Mental Force* (2002).

26 This is also in accord with the interactive role an observer plays in terms of reinfluencing the quantum physics that are being observed. See David Bohm, *The Undivided Universe* (1993).

27—Even what sorts of questions the mind asks.

28—Interestingly, coherent waves appear to provide more efficient and effective chemical reactions than chaotic wave patterns.

29—Max Jammer, *Einstein and Religion* (1999).

30—Paul Davies, *God and the New Physics* (1983).

31—Pim van Lommel of the Hospital Rijnstate in the Netherlands, in *The Lancet* 12/2001, *Toronto Star* 2/2002, and *Washington Post* 2/2002.

32—Joseph LeDoux, in *Sydney Morning Herald* 1/2003.

RECOMMENDED FURTHER READING

Becker, Robert	*Cross Currents*	Tarcher Putnam	(1990)
Bohm, David	*The Undivided Universe*	Routledge	(1995)
Hunt, Valerie	*Infinite Mind*	Malibu Publishing	(1996)
LeDoux, Joseph	*Synaptic Self*	Viking	(2002)
Nichol, Lee (ed),	*The Essential David Bohm*	Routledge	(2003)
McCraty, Rollin	*The HeartMath Report*	Institute of HeartMath	(2001)
Schwartz, Jeffrey, et al	*The Mind and the Brain*	Regan	(2002)

BIBLIOGRAPHY

Aczel, Amir	*Entanglement*	Four Walls Eight Windows	(2002)
Alper, Matthew	*The God Part of the Brain*	Rogue Press	(2001)
Anonymous	*The Teachings of Buddha*	Kosaido	(1989)
Appleyard, Bryan	*Science and the Soul*	Picador	(1992)
Asimov, Issac	*Atom*	Dutton/Plume	(1992)
Becker, Robert	*The Body Electric*	William Morrow	(1985)
Becker, Robert	*Cross Currents*	Tarcher Putnam	(1990)
Berg, Yehuda	*Power of Kabbalah*	Jodere Group	(2002)
Blood, Casey	*Science, Sense and Soul*	St. Martin's Press	(2001)
Bohm, David	*The Undivided Universe*	Routledge	(1995)
Bohm, David	*Wholeness and the Implicate Order*	Routledge	(1996)
Brennan, Barbara	*Hands of Light*	Bantam DoubleDay Dell	(1988)
Brennan, Barbara	*Light Emerging*	Bantam Books	(1993)
Brody, Howard	*The Placebo Response*	Harper Collins	(2001)
Brown, Lowell	*Quantum Field Theory*	Cambridge University Press	(1993)
Burr, Harold Saxton	*Blueprint for Immortality*	Neville Spearman Pub.	(1972)
Campbell, Joseph	*Thou Art That*	New World Library	(2001)
Chia, Mantak & Maneewan	*Awaken Healing Light of the Tao*	Healing Tao Books	(1993)
Chalmers, David	*The Conscious Mind*	Oxford University Press	(1997)
Chopra, Deepak	*Quantum Healing*	Bantam New Age	(1989)

Conlan, Roberta, et al	*States of Mind*	John Wiley & Sons	(2001)
Conte, Rolland, et al	*Theory of High Dilutions*	Dynasol Ltd	(1997)
Cousins, Norman	*Anatomy of an Illness*	Bantam Books	(1981)
Crick, Francis	*Astonishing Hypothesis: Scientific Search for the Soul*	Simon & Schuster	(1995)
Dalai Lama	*A Simple Path*	Thorsons	(1997)
Davies, Paul	*God and the New Physics*	Touchstone S&S	(1983)
Davies, Paul	*The Mind of God*	Simon & Schuster	(1992)
Delgado, Jose	*Physical Control of the Mind*	Harper & Row	(1969)
Dennett, Daniel	*Consciousness Explained*	Little, Brown & Co.	(1992)
Diener, Ed	*Culture and Subjective Well-being*	MIT Press	(2000)
Edelman, Gerald, et al	*The Universe of Consciousness*	Basic Books	(2001)
Edinger, Edward	*Science of the Soul: A Jungian Perspective*	Inner City Books	(2002)
Eliade, Mircea	*World Religions*	Harper San Francisco	(1991)
Feynman, Richard	*QED (Quantum Electrodynamics)*	Princeton University Press	(1988)
Filkin, David	*Stephen Hawking's Universe*	BBC	(1997)
Fowler, James	*Stages of Faith*	Harper San Francisco	(1995)
Frost, Nina, et al	*Soul Mapping*	Avalon Press	(2000)
Gamliel, Dan, et al	*Stochastic Processes in Magnetic Resonance*	World Scientific Pub.	(1995)
Gerber, Richard	*Vibrational Medicine*	Bear & Co.	(2001)
Gershon, Michael	*The Second Brain*	Harper Collins	(1999)
Greene, Brian	*The Elegant Universe*	Norton	(1999)
Greenfield, Susan	*Journey to the Centers of the Mind*	WH Freeman Co	(1995)

Greenfield, Susan	*The Private Life of the Brain*	John Wiley & Sons	(2001)
Hawking, Stephen	*A Brief History of Time*	Bantam	(1987)
Hawking, Stephen	*The Universe in a Nutshell*	Bantam	(2001)
Hay, Louise	*The Power is Within You*	Hay House	(1991)
Henbest, Nigel, et al	*The New Astronomy*	Cambridge University	(1985)
Hunt, Valerie	*Infinite Mind*	Malibu Publishing	(1996)
Jammer, Max	*Einstein and Religion*	Princeton University Press	(2000)
Joseph, Rhawn	*Transmitter to God: The Limbic System*	University Press CA	(2001)
Joseph, Rhawn (ed), et al	*Neurotheology*	University Press CA.	(2002)
Jung, Carl	*Memories, Dreams, Reflections*	Flamingo	(1983)
Kasser, Tim,	*The High Price of Materialism*	MIT Press	(2002)
Lederman, Leon, et al	*From Quarks To Cosmos*	WH Freeman	(1989)
LeDoux, Joseph	*Synaptic Self*	Viking	(2002)
McCraty, Rollin	*The HeartMath Report*	Institute of HearthMath	(2001)
Mesko, Sabrina	*Healing Mudras*	Ballantine Books	(2000)
Moody, Harry, et al	*The Five Stages of the Soul*	Anchor Books	(1997)
Moore, Thomas	*The Soul's Religion*	Harper Collins	(2002)
Myers, David	*The Pursuit of Happiness*	William Morrow & Co.	(1993)
Newberg, Andrew, et al	*Why God Won't Go Away*	Ballantine	(2001)
Nichol, Lee (ed),	*The Essential David Bohm*	Routledge	(2003)
O'Keefe, John, et al,	*The Hippocampus as a Cognitive Map*	Oxford University Press	(1978)
Oschman, James	*Energy Medicine: The Scientific Basis*	Churchill Livingstone	(2000)

Pagels, Elaine	*The Gnostic Gospels*	Penguin	(1979)
Penfield, Wilder	*Mystery of the Mind*	Princeton University Press	(1975)
Penrose, Roger, et al	*The Large, the Small and the Human Mind*	Cambridge University Press	(2000)
Penrose, Roger, et al	*The Emperor's New Mind*	Oxford University Press	(2002)
Persinger, Michael	*ELF & VLF Electromagnetic Field Effects*	Perseus	(1974)
Persinger, Michael	*Neurophysiological Basis of God Beliefs*	Greenwood Publishing	(1987)
Persinger, Michael	*The Paranormal*	Irvington Publishers	(1974)
Persinger, Michael	*The Weather Matrix & Human Behavior*	Greenwood Publishing	(1980)
Persinger, Michael, et al	*Space-Time Transients & Unusual Events*	Nelson Hall	(1977)
Pert, Candace	*Molecules of Emotion*	Touchstone S&S	(1997)
Rumi	*Essential Rumi*	Coleman	(1995)
Ryder, Lewis	*Quantum Field Theory*	Cambridge University Press	(1996)
Satinover, Jeffrey	*The Quantum Brain*	John Wiley & Sons	(2002)
Schwarz, Albert	*Quantum Field Theory*	Springer-Verlag	(1993)
Schwartz, Jeffrey, et al	*The Mind and the Brain*	Regan	(2002)
Seligman, Martin	*Authentic Happiness*	The Free Press	(2002)
Shealy, Norman	*Encyclopedia of Healing Remedies*	Element	(1998)
Shealy, Norman	*Healing Remedies*	Element	(1998)
Siblerud, Robert	*The Science of the Soul*	New Science Publications	(1998)
Smolin, Lee	*Three Roads to Quantum Gravity*	Basic Books	(2001)
Stannard, Bruce	*The God Experiment*	Hidden Spring	(2000)
Steiner, Rudolf	*How to Know Higher Worlds*	Anthroposophical	(1995)

Steiner, Rudolf	*Freud, Jung and Spiritual Psychology*	Anthroposophical	(1990)
Tanlsey, David	*Radionics and the Subtle Anatomy of Man*	CW Daniel & Co.	(1972)
Taylor, John	*Hidden Unity in Nature's Laws*	Cambridge University Press	(2001)
Too, Lillian	*Feng Shui*	Barnes & Noble Books	(1999)
Unknown	*Kabbalah*	Various/Bantam	(1995)
Unknown	*Koran*	Penguin Classics	(2000)
Unknown	*Tibetan Book of the Dead*	Bantam	(1994)
Unknown	*Way of the Pilgrim*	Harper San Francisco	(1983)
Various	*Bhagavad Gita*	Various/Bantam	(1998)
Various	*Tao Te Ching*	Various/Bantam	(1990)
Wilson, Ian	*Jesus: the Evidence*	Weidenfeld & Nicholson	(1984)
Wise, Anna	*High Performance Mind*	Jeremy Press Tarcher/Putnam	(1995)
Zee, Anthony	*Quantum Field Theory in A Nutshell*	Princeton University Press	(2003)

0-595-28418-3